MW00814765

Photoacoustic Tomography (PAT)

Photoacoustic Tomography (PAT)

Editors

Xueding Wang
Xinmai Yang
Xose Luis Dean-Ben

MDPI • Basel • Beijing • Wuhan • Barcelona • Belgrade • Manchester • Tokyo • Cluj • Tianjin

Editors

Xueding Wang
University Michigan Ann Arbor,
Department of Biomedical
Engineering, Ann Arbor
USA

Xinmai Yang
Department of
Mechanical Engineering,
University of Kansas
USA

Xose Luis Dean-Ben
Institute for Biological and
Medical Imaging,
Helmholtz Zentrum Munich &
Technical University of Munich
Germany

Editorial Office
MDPI
St. Alban-Anlage 66
4052 Basel, Switzerland

This is a reprint of articles from the Special Issue published online in the open access journal *Applied Sciences* (ISSN 2076-3417) (available at: https://www.mdpi.com/journal/applsci/special_issues/Photoacoustic_Tomography_PAT).

For citation purposes, cite each article independently as indicated on the article page online and as indicated below:

LastName, A.A.; LastName, B.B.; LastName, C.C. Article Title. *Journal Name* **Year**, *Article Number*, Page Range.

ISBN 978-3-03943-643-9 (Hbk)
ISBN 978-3-03943-644-6 (PDF)

Contents

About the Editors

Xueding Wang Optical Imaging Laboratory at the U-M School of Medicine, is focused on imaging system development, and the adaptation of novel diagnostic and therapeutic technologies to laboratory research and clinical settings, especially those involving light and ultrasound. A major part of our research is focused on the clinical applications of photoacoustic imaging, including those involving breast cancer, inflammatory arthritis, prostate cancer, liver conditions, bowel disease, eye conditions and brain disorders. We are also interested in the design and fabrication of multi-functional nanoparticle agents (e.g., metallic, hydrogel) for both diagnosis and therapy, as well as interactions between nanoparticles and cells.

Xinmai Yang is currently an associate professor at Department of Mechanical Engineering and Institute for Bioengineering Research at the University of Kansas. He is an expert in acoustic/optical imaging and therapy with an extensive background in biomedical ultrasound and photoacoustic imaging. His research extends from linear acoustics to non-linear acoustics and laser/ultrasound interaction, including acoustic cavitation, high-intensity focused ultrasound, and photoacoustic imaging. His current research interest focuses on combining light and ultrasound for imaging and therapy. His primary contributions are in the areas of cavitation bubble dynamics in soft tissue under high-intensity ultrasound, photoacoustic brain imaging, and photo-mediated ultrasound therapy.

Xose Luis Dean-Ben received his Automatics and Electronics Engineering and PhD degrees from the University of Vigo in 2004 and 2009, respectively. Since 2010, he has been working in the field of optoacoustic (photoacoustic) imaging. He contributed to the development of both new systems and processing algorithms, as well as to the demonstration of new bio-medical applications in cancer, cardiovascular biology and neuroscience. Currently, he serves as a senior scientist and group leader at the University of Zürich and ETH Zürich. He has co-authored more than 60 papers in peer-reviewed journals on optoacoustic imaging.

Editorial

Special Issue on Photoacoustic Tomography

Xueding Wang [1,*], Xinmai Yang [2,*] and Xose Luis Dean-Ben [3]

1 Department of Biomedical Engineering, University of Michigan, Ann Arbor, MI 48105, USA
2 Department of Mechanical Engineering, University of Kansas, Lawrence, KS 66045, USA
3 Institute for Biological and Medical Imaging, Helmholtz Zentrum Munich & Technical University of Munich, 85748 Garching, Germany; xl.deanben@pharma.uzh.ch
* Correspondence: xdwang@umich.edu (X.W.); xmyang@ku.edu (X.Y.)

Received: 24 September 2019; Accepted: 30 September 2019; Published: 8 October 2019

Biomedical photoacoustic (or optoacoustic) tomography (PAT), or more generally, photoacoustic imaging (PAI), has been an active area of study and development in the last two decades [1,2]. As an emerging hybrid imaging modality that combines the high-contrast of optical imaging with the high spatial resolution of ultrasound imaging, PAI has been demonstrated to have broad applications in areas including basic research, pre-clinical investigation and clinical studies [3]. It offers great specificity with the ability to detect light-absorbing chomophores, such as hemoglobin, melanin, lipids, etc. This ability enables PAI to provide rich soft-tissue information regarding anatomy and function. The application of contrast agents further enables PAI to provide molecular information [4]. Because of these advantages, PAI has great potential in the clinical diagnosis of various diseases, such as cancer, stroke, atherosclerosis, arthritis, etc.

The goal of the current Special Issue is to showcase some of the most recent research and development into this fascinating imaging technology. Indeed, there are so many aspects that an engineer can work on to improve PAI and facilitate its translation into the clinic, from ultrasonic detector design, light delivery and illumination, to reconstruction algorithms. The 11 articles in the current Special Issue represent some of these efforts from research groups all over the world. Alijabbari et al. [5] were able to demonstrate that a full-ring system, not only for the ultrasound transducer, but also for light illumination, can provide superior imaging quality because of the improved illumination configuration. A similar idea was also adopted by Avanaki's research group [6], as well as by Sun et al. [7]. However, these research groups presented a low-cost system made of multi single-element transducers, rather than an expensive array system that could be cost-prohibitive for many users. Improvement on the reconstruction algorithm is another major theme for the current Special Issue. Various methods were presented to improve the quality of PAT images by overcoming artifacts due to respiration [8], reflection [9] and scattering [10]. Novel approaches were proposed to reconstruct PAT images based on full-field detection [11] as well as detections with limited view angles [12]. Based on deconvolution and empirical mode decomposition, Guo et al. was able to improve final image quality by alleviating signal aliasing introduced by N-shape waves [13]. Ma et al. introduced a machine-learning method to PAT image analysis in order to measure the size of adipocytes [14]. Finally, Kothapalli and his research group developed a single simulation platform for both photoacoustic and thermoacoustic imaging [15]. These are truly some of the exciting developments.

PAI is a fast-evolving research and development field. Researchers all over the world are working diligently on continuously improving the technology and translating it to the clinic environment. The current Special Issue, with 11 articles from research groups across different parts of the world, is a very small sample to demonstrate the potential impact of this technology. We hope you will enjoy reading these selected articles and look forward to your future contribution to a future Special Issue.

References

1. Beard, P. Biomedical photoacoustic imaging. *Interface Focus* **2011**, *1*, 602–631. [CrossRef] [PubMed]
2. Xia, J.; Yao, J.; Wang, L.V. Photoacoustic tomography: Principles and advances. *Electromagn. Waves* **2014**, *147*, 1–22. [CrossRef]
3. Steinberg, I.; Huland, D.M.; Vermesh, O.; Frostig, H.E.; Tummers, W.S.; Gambhir, S.S. Photoacoustic clinical imaging. *Photoacoustics* **2019**, *14*, 77–98. [CrossRef] [PubMed]
4. Zeng, L.; Ma, G.; Lin, J.; Huang, P. Photoacoustic Probes for Molecular Detection: Recent Advances and Perspectives. *Small* **2018**, *14*, e1800782. [CrossRef]
5. Alijabbari, N.; Alshahrani, S.S.; Pattyn, A.; Mehrmohammadi, M. Photoacoustic Tomography with a Ring Ultrasound Transducer: A Comparison of Different Illumination Strategies. *Appl. Sci.* **2019**, *9*, 3094. [CrossRef]
6. Zafar, M.; Kratkiewicz, K.; Manwar, R.; Avanaki, M. Development of Low-Cost Fast Photoacoustic Computed Tomography: System Characterization and Phantom Study. *Appl. Sci.* **2019**, *9*, 374. [CrossRef]
7. Sun, M.J.; Hu, D.P.; Zhou, W.X.; Liu, Y.; Qu, Y.W.; Ma, L.Y. 3D Photoacoustic Tomography System Based on Full-View Illumination and Ultrasound Detection. *Appl. Sci.* **2019**, *9*, 1904. [CrossRef]
8. Ron, A.; Davoudi, N.; Dean-Ben, X.L.; Razansky, D. Self-Gated Respiratory Motion Rejection for Optoacoustic Tomography. *Appl. Sci.* **2019**, *9*, 2737. [CrossRef]
9. Shan, H.M.; Wang, G.; Yang, Y. Accelerated Correction of Reflection Artifacts by Deep Neural Networks in Photo-Acoustic Tomography. *Appl. Sci.* **2019**, *9*, 2615. [CrossRef]
10. Rui, W.; Liu, Z.P.; Tao, C.; Liu, X.J. Reconstruction of Photoacoustic Tomography Inside a Scattering Layer Using a Matrix Filtering Method. *Appl. Sci.* **2019**, *9*, 2071. [CrossRef]
11. Zangerl, G.; Haltmeier, M.; Nguyen, L.V.; Nuster, R. Full Field Inversion in Photoacoustic Tomography with Variable Sound Speed. *Appl. Sci.* **2019**, *9*, 1563. [CrossRef]
12. Omidi, P.; Zafar, M.; Mozaffarzadeh, M.; Hariri, A.; Haung, X.Z.; Orooji, M.; Nasiriavanaki, M. A Novel Dictionary-Based Image Reconstruction for Photoacoustic Computed Tomography. *Appl. Sci.* **2018**, *8*, 1570. [CrossRef]
13. Guo, C.W.; Chen, Y.N.; Yuan, J.; Zhu, Y.H.; Cheng, Q.; Wang, X.D. Biomedical Photoacoustic Imaging Optimization with Deconvolution and EMD Reconstruction. *Appl. Sci.* **2018**, *8*, 2113. [CrossRef]
14. Ma, X.; Cao, M.; Shen, Q.H.; Yuan, J.; Feng, T.; Cheng, Q.; Wang, X.D.; Washabaugh, A.R.; Baker, N.A.; Lumeng, C.N.; et al. Adipocyte Size Evaluation Based on Photoacoustic Spectral Analysis Combined with Deep Learning Method. *Appl. Sci.* **2018**, *8*, 2178. [CrossRef]
15. Fadden, C.; Kothapalli, S.R. A Single Simulation Platform for Hybrid Photoacoustic and RF-Acoustic Computed Tomography. *Appl. Sci.* **2018**, *8*, 1568. [CrossRef] [PubMed]

Article

Photoacoustic Tomography with a Ring Ultrasound Transducer: A Comparison of Different Illumination Strategies

Naser Alijabbari [1,†]**, Suhail S. Alshahrani** [1,†]**, Alexander Pattyn** [1] **and Mohammad Mehrmohammadi** [1,2,3,*]

1   Department of Biomedical Engineering, Wayne State University, Detroit, MI 48201, USA
2   Department of Electrical and Computer Engineering, Wayne State University, Detroit, MI 48201, USA
3   Barbara Ann Karmanos Cancer Institute, Detroit, MI 48201, USA
*   Correspondence: mehr@wayne.edu
†   Those authors are contributed equally to this work.

Received: 3 June 2019; Accepted: 27 July 2019; Published: 31 July 2019

Featured Application: Biological diagnostic applications based on endogenous or exogenous chromophores and early breast cancer detection in dense tissue.

Abstract: Photoacoustic (PA) imaging is a methodology that uses the absorption of short laser pulses by endogenous or exogenous chromophores within human tissue, and the subsequent generation of acoustic waves acquired by an ultrasound (US) transducer, to form an image that can provide functional and molecular information. Amongst the various types of PA imaging, PA tomography (PAT) has been proposed for imaging pathologies such as breast cancer. However, the main challenge for PAT imaging is the deliverance of sufficient light energy horizontally through an imaging cross-section as well as vertically. In this study, three different illumination methods are compared for a full-ring ultrasound (US) PAT system. The three distinct illumination setups are full-ring, diffused-beam, and point source illumination. The full-ring system utilizes a cone mirror and parabolic reflector to create the ringed-shaped beam for PAT, while the diffuse scheme uses a light diffuser to expand the beam, which illuminates tissue-mimicking phantoms. The results indicate that the full-ring illumination is capable of providing a more uniform fluence irrespective of the vertical depth of the imaged cross-section, while the point source and diffused illumination methods provide a higher fluence at regions closer to the point of entry, which diminishes with depth. In addition, a set of experiments was conducted to determine the optimum position of ring-illumination with respect to the position of the acoustic detectors to achieve the highest signal-to-noise ratio.

Keywords: full-ring illumination; diffused-beam illumination; point source illumination; ultrasound tomography (UST); photoacoustic tomography (PAT)

---

1. Introduction

Breast cancer is a significant health problem not only in the United States but globally and was the second leading cause of cancer-related death in the United States in 2018 [1]. Mammography, MRI (Magnetic Resonance Imaging), and B-mode ultrasound are the three most common imaging modalities used for breast cancer screening [2,3]. However, each of these modalities has its own unique shortcomings. The sensitivity of mammography in detecting breast lesions decreases in women with high-density breast tissue, and high-density breasts are considered to be more at risk for developing breast cancer [3,4]. In dense breasts, MRI can be used in conjunction with breast mammography to detect breast tumors [5,6]. Nevertheless, the operational cost and availability of MRI imaging limit the accessibility of this modality. Conventional B-mode ultrasound (US) is a high-sensitivity, non-ionizing, and low-cost tool that is widely used for screening various types of human tissues [7,8]. However, false positives due to ultrasound screening result in many unnecessary biopsies [9,10]. Therefore, a more effective breast cancer screening tool is sought.

Photoacoustic tomography (PAT) is an imaging methodology that uses light absorption by endogenous or exogenous chromophores, and subsequent US pressure wave generation for imaging. Photoacoustic (PA) imaging has been demonstrated to be useful for a variety of medical and biological diagnostic applications, including early cancer detection [11–14]. Biomarkers such as vascularity and hypoxia have been shown to have diagnostic value in the differential diagnosis of various types of cancers including breast cancer [15–17]. In addition, when PA is augmented with nano-sized contrast agents, it can provide a reliable platform for the molecular imaging of cancer and its sub-types [18–21]. In clinical applications, PA imaging has been shown to produce real-time molecular and functional information with high resolution at relevant depths [19,22].

Several PAT imaging systems with different illumination and acquisition modes have been developed for breast cancer imaging. However, the observed limitations of these systems are in part due to the non-optimum acoustic signal acquisition or illumination methodologies used. Our presented system is meant to be non-invasive (i.e., both illumination and acquisition are external), with the illumination and measurement system external to the imaged tissue. Our method images a cross-section inside the cylindrical US transducer array by illuminating the targeted area using a ring beam. Therefore, the light has to only diffuse half the tissue diameter that is encountered when using side illumination. Point or diffuse illuminations are suitable for imaging cross-sections close to the point of light entry [11,23–27], and the given fluence drops with light propagating through the tissue towards higher vertical depths as shown in Figure 1. This could make it difficult to access areas close to the chest well.

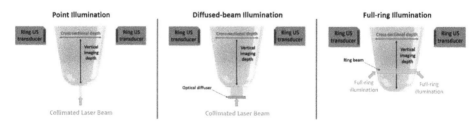

Figure 1. The three methods of illumination for photoacoustic tomography (PAT) imaging that are compared in this study, with the definitions of vertical and cross-sectional imaging depths.

Other PAT imaging methods, such as the PA mammoscopy system [28], compress the tissue for better light penetration but can cause discomfort, or a loss of important PA biomarkers arises from the presence of blood by pushing the blood out of the tissue. One type of full-ring illumination system uses an acoustically penetrable optical reflector (APOR). However, APORs can only support low laser energies and US transmission through the reflector is highly angle-dependent [29–31]. Other illumination methods for deep tissue illumination include internal illuminations [32,33]. However, internal illumination is difficult to develop for breast imaging applications. Therefore, it is vital to develop an alternative solution for improving the uniformity of energy distribution within the breast tissue for more accurate PAT imaging.

Ultrasound tomography (UST) using a ring-shaped US transducer has shown promising results in breast cancer screening [34–38]. In this work, PAT imaging is combined with this novel full ring UST system. The PA imaging modality can be easily combined with UST since both modalities share the same acquisition hardware. For this reason, the addition of the PAT to the UST is straightforward and will provide valuable functional information about a given tissue and is expected to improve the diagnostic capability of breast US for physicians.

The design of our combined UST/PAT imaging system has been previously presented [39–42]. This setup uses a ring illumination in conjunction with a ring US transducer for combined UST/PAT imaging. The ring-shaped beam in this system is generated by using a cone mirror and a parabolic reflector. This work specifically compares three different illumination methodologies for PAT imaging: full-ring, diffuse, and point illuminations. Using new findings from the three methods, it aims to show that full-ring illumination is the most effective method for creating PAT images due to its inherent cross-sectional fluence uniformity across vertical imaging depths (Figure 1). This is especially important for breast cancer screening when imaging close to the chest wall proves difficult.

The three illumination methods are compared by imaging a three-layer polyvinyl chloride (PVC) tissue-mimicking phantom to gauge the advantages and disadvantages of common PAT imaging techniques. The experiments presented in this paper also all use the same data acquisition system and settings. Comparisons are made between PAT amplitudes for each cross-section and illumination methodology. Furthermore, the optimum position of the ring beam with respect to the targeted cross-section is examined.

2. Material and Methods

2.1. UST/PAT Acquisition System

A 200 mm diameter, 256-element ring US transducer (Analogic Corporation, Canada) with a center frequency of 2 MHz and bandwidth of 60% was used for all data acquisition. The presented system has a measured lateral resolution of 1 mm as determined by measuring a 200 micrometer light-absorbing string. This transducer has an element pitch of 2.45 mm and a height of 9 mm. The scattered US waves from a PA imaging event are recorded by all 256 elements using a sampling frequency of 8.33 MHz. As shown in Figure 2a, the US ring transducer is housed in an acrylic tank and is supplied with degassed, distilled water. During PAT imaging, the ring US transducer uses a 10 dB linear, time gain compensation (TGC) for acquiring the data, which is designed to optimize the signal-to-noise ratio (SNR) for the given phantom.

2.2. Laser Source and Light Illumination Schemes

A tunable, 10 nanoseconds pulsed laser (Phocus Core, Optotek, Carlsbad, CA, USA) was used for all PAT imaging experiments. This laser generates around 100 mJ per pulse at 532 nm. In the full-ring illumination mode, a large parabolic reflector (P19-0300, Optiforms Inc., Temecula, CA, USA) was used with a 10 mm diameter cone mirror (68-791, Edmund Optics, Barrington, NJ, USA) to create the 4 mm ring-shaped beam on the phantom surface (Figure 2a). Since the beam position is stationary, neither the cone mirror nor the parabolic reflector is mobile. The ring location was adjusted across each cross-section by translating the phantom in the vertical direction (Figure 2b). For the diffused-beam experiments, a 120 grit ground glass diffuser (DG10-120, Thorlabs Inc., Newton, NJ, USA) was placed in the laser light path inside the water tank after removing the cone mirror (Figure 2c). Finally, point illumination only uses the 45-degree mirror for directing the laser beam onto the phantom (Figure 2d).

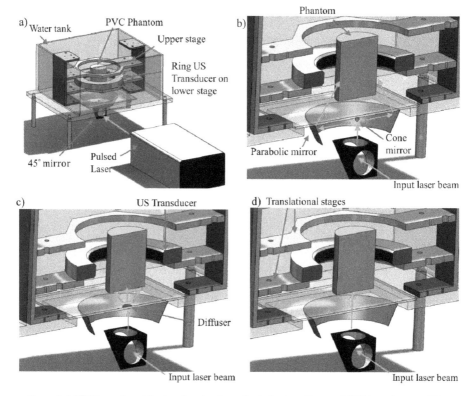

Figure 2. (**a**) PAT experimental setup showing the water tank, ring ultrasound (US) transducer, and the translational stages. The experimental setups for the (**b**) full ring, (**c**) diffuse-beam, and (**d**) point illumination of the phantom.

2.3. Tissue-Mimicking Phantoms

The performance of the three different illumination methods was evaluated with regards to the PA imaging depth using a phantom made of polyvinyl chloride (PVC) (M-F Manufacturing Super Soft, Fort Worth, TX, USA), with 0.2% fine ground silica (US Silica MIN-U-SIL5, Stow, OH, USA) used as an optical diffuser. To create a PVC phantom [43], the plastisol was first mixed with the ground silica and then heated in a microwave to 170 °C. In a mold, three graphite rods with a 2 mm diameter were placed horizontally, and the PVC was poured to create the three layers as shown in Figure 3. After cooling, the phantom was removed from the mold and used for the described experiments. Graphite was chosen as an absorber due to its broadband absorption characteristics and ease of placement inside the phantom.

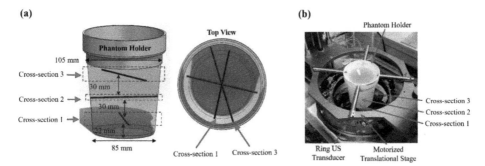

Figure 3. (**a**) Graph illustrates the polyvinyl chloride (PVC) phantom and graphite inclusions and their dimensions. (**b**) A photograph of the experimental setup including the ring US transducer. A motorized translational stage was used to adjust the position of the phantom to acquired images at multiple cross-sections.

2.4. UST and PAT Image Reconstruction

In all instance, the waveform method was used to reconstruct the UST images [44], while filtered back-projection was utilized to reconstruct the PAT images [45]. For PAT back-projection reconstruction, the RF (radio frequency) values for each cross-section were averaged 10 times to increase the overall SNR. In UST mode, a 20 dB linear gain TGC was used for acquiring the US images, while in PAT mode, as previously described, a linear 10 dB TGC was used for data acquisition. For US imaging, the used TGC was optimized to minimize the transducer saturation, which resulted in cleaner images. This was done empirically. For PAT imaging, the value chosen was designed to reduce the signal emanating from other cross-sections from appearing in the cross-section of interest.

3. Results and Discussion

3.1. A Comparison of the Three Different Illumination Methods

The results discussed in this section focus on analyzing the PAT images and PA signal amplitudes from progressively deeper phantom cross-sections using the discussed illumination methodologies. Figure 4 shows the UST and PAT images for the first and third cross-sections, which are separated by 60 mm. The PAT images are masked based on the region of interest (ROI) as determined by the UST image. For each illumination method, the PA amplitude is normalized to the highest value for the method across all cross-sections. For example, for the full-ring illumination method, images were normalized to the highest amplitude of PA detected within the three slices, which are separated by 30 mm each. This allows for visualization of the effect of depth on the PA signal amplitudes for each illumination method. As can be seen in the PAT images, the graphite absorber is visible in the first and third layers (i.e., larger vertical depth) of the full-ring illumination method, which is not the case for the diffuse and point-source methods.

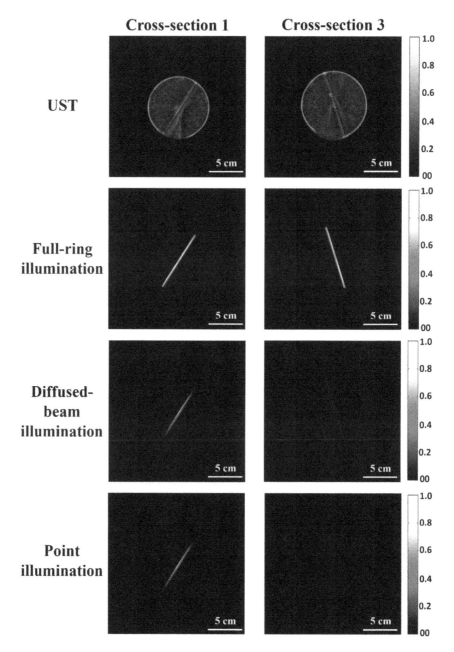

Figure 4. Ultrasound tomography (UST) and normalized PAT images of the PVC phantom with the graphite absorber using the three different illumination techniques for cross-sections 1 and 3.

To further quantify the results shown in Figure 4, the PA signal amplitude across the graphite absorber for each illumination method, and for each cross-section, is plotted in Figure 5. For the full-ring illumination method, the PA values are nearly constant across the three cross-sections (Figure 5a). However, the peak amplitude value of the PA signal decreases by 25 times for point-source illumination and 15 times for the diffused-beam illumination between the first and third cross-sections. As seen by the uniformity in the amplitudes between three cross-sections for the full-ring illumination method, this imaging technique can provide a consistent image regardless of vertical depth. On the other hand, the diffused and point illumination methodologies show variance in PA amplitude signals. This finding demonstrates that the full-ring illumination is capable of providing sufficient fluences at lower vertical depths, which results in detectable and reliable PAT images across cross-sections.

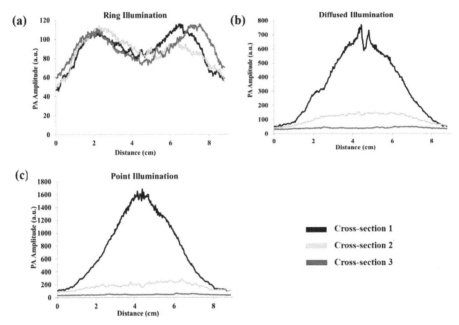

Figure 5. PA amplitude across the graphite absorber for three different cross-sections for: (**a**) full-ring, (**b**) diffused-beam, and (**c**) point illumination.

To further compare the performance of the three illumination strategies, the SNR and contrast-to-noise ratio (CNR) of the PAT images were measured. In the SNR, the value is calculated by:

$$SNR = 20 \times log10\left(\frac{M_S}{M_B}\right)$$

where M_S refers to the mean of the PA signal, as marked by the US image region of interest, while M_B refers to the mean of the phantom background. The CNR is determined by:

$$CNR = 20 \times log10\left(\frac{M_S - M_B}{\sigma_B}\right)$$

where M_S refers to the mean of the PA signal, M_B refers to the mean of the phantom background, and σ_B is the standard deviation of the phantom background. For the full-ring illumination method, the used PA amplitude values are from irradiating the target cross-section 15 mm below the cross-section of interest. As can be seen in Figure 6a,b, the SNR and CNR are nearly constant for the full-ring illumination across the three cross-sections, which is not the case for the diffuse and point illuminations. Figure 6c also plots the amplitudes across the graphite object at the third cross-section for all illumination methods. Based on the laser beam diameter of 8 mm; optical losses in the system; and 100 mJ input energy, the diffuse illumination method has a fluence of 9.3 mJ/cm^2, compared to 175 mJ/cm^2 for point illumination, and 7 mJ/cm^2 for the full ring illumination. Full-ring calculations use a beam height of 4 mm circumferentially on a 9 cm diameter phantom, and diffused beam calculations use a beam diameter of 30 mm at the phantom. Even though point illumination has about 25 times the fluence of the full-ring illumination method, its amplitude is much smaller at this cross-sectional depth.

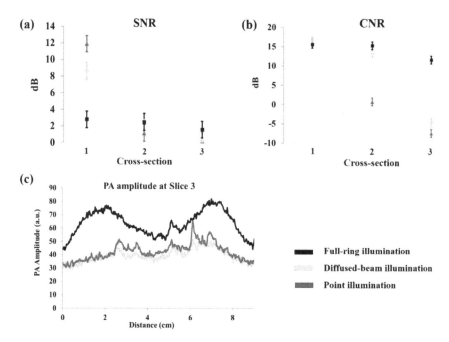

Figure 6. The signal-to-noise ratio (SNR) and contrast-to-noise ratio (CNR) of the PA amplitudes at three different cross-sections are plotted in (**a**,**b**), respectively. For the full-ring illumination, the values were determined based on the illumination at 15 mm below the cross-section of interest. (**c**) Plots the PA amplitude for the top cross-section (cross-section 3) for full-ring, diffuse, and point illumination.

When compared to the diffused-beam and point illuminations, full-ring illumination has a higher PA amplitude at Cross-section 3. It has a near constant SNR, CNR, and PA signal amplitude across all cross-sections, making it an effective illumination method for breast imaging. Given that imaging breast regions close to the chest wall (i.e., large vertical depth) are clinically important, the full-ring illumination shows promise in accessing these regions and thus provides a means for reliable whole breast PAT imaging with a ring US transducer.

3.2. The PA Amplitude of the Targeted Cross-Section as a Function of Illumination Position

The optimum position for the full-ring beam was investigated by evaluating the effect of the distance between the ring-shaped beam and the targeted cross-section. The distance between the targeted cross-section and the full-ring beam was changed within a range of 0–20 mm. Zero millimeters represents the case where the ring beam is illuminating the targeted cross-section at the graphite rod, while the 20 mm case is when the ring beam is 20 mm below the graphite rod (Figure 7a). In this study, a selected cross-section was imaged while changing the illumination location from 0 to 20 mm. The five different positions chosen to illuminate the targeted cross-section of the graphite rod were 0, 5, 10, 15, and 20 mm. A 532 nm laser with 100 mJ per pulse energy was used for this experiment. The location of the full-ring illumination was adjusted by translating the phantom in steps of 5 mm in the vertical direction.

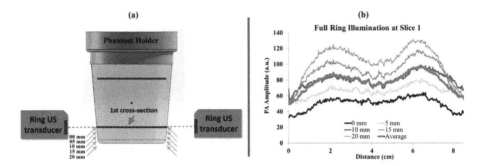

Figure 7. (a) The image shows the different positions of the ring beam based on the targeted cross-section (Cross-section 1). The targeted cross-section is located in the central field of view of the US elements. (b) PA amplitude at Cross-section 1 plotted as a function of illumination depth below the cross-section.

In Figure 7b, the PA amplitude was measured by drawing a line across the targeted graphite for all five positions. The overall shape of the PA amplitude is constant while the maximum increases as one moves further below the desired imaging cross-section. This increased visibility could be due to the fact that the incidence angle of the beam illuminates the central part of the object more directly as it moves below the cross-section. A possible reason for the stronger PA signal, when illuminating 20 mm below the targeted cross-section, could the incident angle of the light diffusion within the tissue. In the current prototype, the ring mirror is illuminating the object cross-section at an angle of 39 degrees with respect to the object surface. The optimum illumination will occur if the ring beam falls normal to the surface. The 0 and 5 mm cross-sections use the US image as a mask to crop out the large PA peaks generated at the surface of the phantom. The graphite absorber was embedded within the PVC background with a margin from the surface. Here, we only evaluated the signal arising from the absorber. In cases where the illumination was coincident with the center of the transducer, a strong PA signal from the surface was observed (not shown in these masked images). Hence, imaging below the transducer can help to better visualize more central parts of the object due to elimination of a large PA signal arising from the light-entering surface.

The large PA signal at the surface is primarily due to the large fluence at the surface, which can affect the visualization of deeper regions due to a limited dynamic range in PA acquisition. This is not the case where the illumination was adjusted to 10 to 20 mm with respect to the center of the US detectors (i.e., illuminating below the imaged plane). The averaged PA amplitude from all five illumination methods (shown in red in Figure 7b) is similar to the 10 mm illumination results. It is worth mentioning that these results are not necessarily general for all ring illumination systems and are dependent on the incident angle of the beam with respect to the object (39 degrees in our case). The results might vary in other ring illumination systems if the incident angle is changed.

These findings are important because they help to define the illumination scenario based on the characteristics and size of the given object. For example, for a large diameter phantom one might need to acquire PAT images using more than one position of full-ring illumination to cover the regions closer to the outer surface and deeper regions (with illumination offset from the imaging plane). In addition, illuminating below the targeted cross-section (below the central line of the US elements) could help to limit the high PA signals that are generated from the outer surface of the scanned object. The used illumination method has a significant effect on the imaging's vertical depth, which is significant when visualizing anatomy such as the breast.

4. Conclusions

Phantoms with horizontal graphite absorbers were used to examine the efficacy of full-ring, diffuse, and point illumination PAT imaging using a ring US transducer. The full-ring illumination was able to provide a more uniform fluence irrespective of the vertical depth of the imaged cross-section, while the point source and diffused-beam resulted in a high fluence at the point of entry, which diminished with vertical depth. In addition, the preliminary results indicate that illuminating an object 10–15 mm below the imaged cross-section of interest might be optimal for this system since it avoids large surface reflections and provides better coverage to the cross-section.

Author Contributions: N.A. and S.S.A. contributed to: conceptualization, methodology, formal analysis, investigation, data curation, writing—original draft preparation, writing—review and editing, and visualization. A.P. developed the software used for creating the PAT back-projection images. M.M. was involved in conceptualization, methodology, formal analysis, investigation, writing, the review and editing, supervision, and project administration.

Funding: This project was supported by Department of Defense (Breast Cancer Research program, Award number (W81XWH-18-1-0039). Naser Alijabbari (NA) was further supported by Office of the Vice President for Research (OVPR) offers the Faculty Competition for Postdoctoral Fellows.

Acknowledgments: The authors acknowledge Wayne State University for their financial support and the Barbara Ann Karmanos Cancer Institute. Also, we would like to acknowledge Neb Duric from Karmanos Cancer Institute for his guidance and assistance with using the UST system. MM was supported by Department of Defense (Breast Cancer Research program, Award number (W81XWH-18-1-0039), University Research Grant from Wayne State University Provost office, Technology Development Incubator Award through Wayne State University, and pilot research funding from Barbara Ann Karmanos Cancer Institute.

Conflicts of Interest: The authors have no relevant financial interests in the manuscript and no other potential conflicts of interest to disclose.

References

1. Siegel, R.L.; Miller, K.D.; Jemal, A. Cancer statistics, 2019. *CA A Cancer J. Clin.* **2019**, *69*, 7–34. [CrossRef] [PubMed]

2. Elmore, J.G.; Armstrong, K.; Lehman, C.D.; Fletcher, S.W. Screening for breast cancer. *JAMA* **2005**, *293*, 1245–1256. [CrossRef] [PubMed]

3. Kuhl, C.K.; Schrading, S.; Leutner, C.C.; Wardelmann, E.; Fimmers, R.; Schild, H.H.; Morakkabati-Spitz, N.; Kuhn, W. Mammography, Breast Ultrasound, and Magnetic Resonance Imaging for Surveillance of Women at High Familial Risk for Breast Cancer. *J. Clin. Oncol.* **2005**, *23*, 8469–8476. [CrossRef] [PubMed]

4. Tamimi, R.M.; Byrne, C.; Colditz, G.A.; Hankinson, S.E. Endogenous Hormone Levels, Mammographic Density, and Subsequent Risk of Breast Cancer in Postmenopausal Women. *J. Natl. Cancer Inst.* **2007**, *99*, 1178–1187. [CrossRef] [PubMed]

5. Saslow, D.; Boetes, C.; Burke, W.; Harms, S.; Leach, M.O.; Lehman, C.D.; Morris, E.; Pisano, E.; Schnall, M.; Sener, S.; et al. American Cancer Society Guidelines for Breast Screening with MRI as an Adjunct to Mammography. *Obstet. Gynecol. Surv.* **2007**, *62*, 458–460. [CrossRef]

6. Lee, C.H.; Dershaw, D.D.; Kopans, D.; Evans, P.; Monsees, B.; Monticciolo, D.; Brenner, R.J.; Bassett, L.; Berg, W.; Feig, S.; et al. Breast Cancer Screening With Imaging: Recommendations From the Society of Breast Imaging and the ACR on the Use of Mammography, Breast MRI, Breast Ultrasound, and Other Technologies for the Detection of Clinically Occult Breast Cancer. *J. Am. Coll. Radiol.* **2010**, *7*, 18–27. [CrossRef] [PubMed]

7. Jensen, J.A. Medical ultrasound imaging. *Prog. Biophys. Mol. Biol.* **2007**, *93*, 153–165. [CrossRef] [PubMed]

8. Gordon, P.B.; Goldenberg, S.L. Malignant breast masses detected only by ultrasound. A retrospective review. *Cancer* **1995**, *76*, 626–630. [CrossRef]

9. Madjar, H. Role of Breast Ultrasound for the Detection and Differentiation of Breast Lesions. *Breast Care* **2010**, *5*, 109–114. [CrossRef]

10. Corsetti, V.; Houssami, N.; Ferrari, A.; Ghirardi, M.; Bellarosa, S.; Angelini, O.; Bani, C.; Sardo, P.; Remida, G.; Galligioni, E.; et al. Breast screening with ultrasound in women with mammography-negative dense breasts: Evidence on incremental cancer detection and false positives, and associated cost. *Eur. J. Cancer* **2008**, *44*, 539–544. [CrossRef]

11. Kruger, R.A.; Lam, R.B.; Reinecke, D.R.; Del Rio, S.P.; Doyle, R.P. Photoacoustic angiography of the breast. *Med. Phys.* **2010**, *37*, 6096–6100. [CrossRef] [PubMed]

12. Menke, J. Photoacoustic breast tomography prototypes with reported human applications. *Eur. Radiol.* **2015**, *25*, 2205–2213. [CrossRef] [PubMed]

13. Mallidi, S.; Luke, G.P.; Emelianov, S. Photoacoustic imaging in cancer detection, diagnosis, and treatment guidance. *Trends Biotechnol.* **2011**, *29*, 213–221. [CrossRef] [PubMed]
14. Mehrmohammadi, M.; Yoon, S.J.; Yeager, D.; Emelianov, S.Y. Photoacoustic Imaging for Cancer Detection and Staging. *Curr. Mol. Imaging* **2013**, *2*, 89–105. [CrossRef] [PubMed]
15. Siphanto, R.I.; Thumma, K.K.; Kolkman, R.G.M.; Van Leeuwen, T.G.; De Mul, F.F.M.; Van Neck, J.W.; Van Adrichem, L.N.A.; Steenbergen, W. Serial noninvasive photoacoustic imaging of neovascularization in tumor angiogenesis. *Opt. Express* **2005**, *13*, 89–95. [CrossRef] [PubMed]
16. Brahimi-Horn, M.C.; Chiche, J.; Pouysségur, J. Hypoxia and cancer. *J. Mol. Med.* **2007**, *85*, 1301–1307. [CrossRef] [PubMed]
17. Heijblom, M.; Klaase, J.M.; Engh, F.M.V.D.; Van Leeuwen, T.G.; Steenbergen, W.; Manohar, S. Imaging tumor vascularization for detection and diagnosis of breast cancer. Technol. *Cancer Res. Treat.* **2011**, *10*, 607–623. [CrossRef] [PubMed]
18. Alcantara, D.; Leal, M.P.; García-Bocanegra, I.; García-Martín, M.L. Molecular imaging of breast cancer: Present and future directions. *Front. Chem.* **2014**, *2*, 112. [CrossRef] [PubMed]
19. Luke, G.P.; Yeager, D.; Emelianov, S.Y. Biomedical applications of photoacoustic imaging with exogenous contrast agents. *Ann. Biomed. Eng.* **2012**, *40*, 422–437. [CrossRef] [PubMed]
20. Wang, J.; Jeevarathinam, A.S.; Humphries, K.; Jhunjhunwala, A.; Chen, F.; Hariri, A.; Miller, B.R.; Jokerst, J.V. A Mechanistic Investigation of Methylene Blue and Heparin Interactions and Their Photoacoustic Enhancement. *Bioconjug. Chem.* **2018**, *29*, 3768–3775. [CrossRef] [PubMed]
21. Jeevarathinam, A.S.; Pai, N.; Huang, K.; Hariri, A.; Wang, J.; Bai, Y.; Wang, L.; Hancock, T.; Keys, S.; Penny, W.; et al. A cellulose-based photoacoustic sensor to measure heparin concentration and activity in human blood samples. *Biosens. Bioelectron.* **2019**, *126*, 831–837. [CrossRef] [PubMed]
22. Wang, L.V. Prospects of photoacoustic tomography. *Med. Phys.* **2008**, *35*, 5758–5767. [CrossRef] [PubMed]
23. Ermilov, S.A.; Khamapirad, T.; Conjusteau, A.; Leonard, M.H.; Lacewell, R.; Mehta, K.; Miller, T.; Oraevsky, A.A. Laser optoacoustic imaging system for detection of breast cancer. *J. Biomed. Opt.* **2009**, *14*, 024007. [CrossRef] [PubMed]
24. Kruger, R.A.; Kuzmiak, C.M.; Lam, R.B.; Reinecke, D.R.; Del Rio, S.P.; Steed, D. Dedicated 3D photoacoustic breast imaging. *Med. Phys.* **2013**, *40*, 113301. [CrossRef] [PubMed]
25. Lin, L.; Hu, P.; Shi, J.; Appleton, C.M.; Maslov, K.; Li, L.; Zhang, R.; Wang, L.V. Single-breath-hold photoacoustic computed tomography of the breast. *Nat. Commun.* **2018**, *9*, 2352. [CrossRef] [PubMed]
26. Lin, L.; Hu, P.; Shi, J.; Maslov, K.I.; Appleton, C.M. Clinical photoacoustic computed tomography of the human breast in vivo within a single breath hold. In *Photons Plus Ultrasound: Imaging and Sensing 2018*; International Society for Optics and Photonics: Bellingham, WA, USA, 2018; Volume 10494, p. 104942X.
27. Klosner, M.; Chan, G.; Wu, C.; Heller, D.F.; Su, R.; Ermilov, S.; Brecht, H.P.; Ivanov, V.; Talole, P.; Lou, Y.; et al. Advanced laser system for 3D optoacoustic tomography of the breast. In *Photons Plus Ultrasound: Imaging and Sensing 2016*; International Society for Optics and Photonics: Bellingham, WA, USA, 2016; Volume 9708, p. 97085B.
28. Heijblom, M.; Piras, D.; Xia, W.; Van Hespen, J.C.G.; Engh, F.M.V.D.; Klaase, J.M.; Van Leeuwen, T.G.; Steenbergen, W.; Manohar, S. Imaging breast lesions using the Twente Photoacoustic Mammoscope: Ongoing clinical experience. In *Photons Plus Ultrasound: Imaging and Sensing 2012*; International Society for Optics and Photonics: Bellingham, WA, USA, 2012; Volume 8223, p. 82230C.
29. Deng, Z.; Zhao, H.; Ren, Q.; Li, C. Acoustically penetrable optical reflector for photoacoustic tomography. *J. Biomed. Opt.* **2013**, *18*, 070503. [CrossRef] [PubMed]
30. Deng, Z.; Li, C. Noninvasively measuring oxygen saturation of human finger-joint vessels by multi-transducer functional photoacoustic tomography. *J. Biomed. Opt.* **2016**, *21*, 61009. [CrossRef] [PubMed]
31. Deng, Z.; Li, W. Slip-ring-based multi-transducer photoacoustic tomography system. *Opt. Lett.* **2016**, *41*, 2859–2862. [CrossRef] [PubMed]
32. Li, M.; Lan, B.; Liu, W.; Xia, J.; Yao, J. Internal-illumination photoacoustic computed tomography. *J. Biomed. Opt.* **2018**, *23*, 1–4. [CrossRef]
33. Bungart, B.; Cao, Y.; Yang-Tran, T.; Gorsky, S.; Lan, L.; Roblyer, D.; Koch, M.O.; Cheng, L.; Masterson, T.; Cheng, J.X. Cylindrical illumination with angular coupling for whole-prostate photoacoustic tomography. Biomed. *Opt. Express* **2019**, *10*, 1405–1419. [CrossRef]

34. Duric, N.; Littrup, P.; Schmidt, S.; Li, C.P.; Roy, O.; Bey-Knight, L.; Janer, R.; Kunz, D.; Chen, X.Y.; Goll, J.; et al. Breast imaging with the SoftVue imaging system: First results. In *Medical Imaging 2013: Ultrasonic Imaging, Tomography, and Therapy*; SPIE Medical Imaging; International Society for Optics and Photonics: Bellingham, WA, USA, 2013.

35. Duric, N.; Littrup, P.; Poulo, L.; Babkin, A.; Pevzner, R.; Holsapple, E.; Rama, O.; Glide, C. Detection of breast cancer with ultrasound tomography: First results with the Computed Ultrasound Risk Evaluation (CURE) prototype. *Med. Phys.* **2007**, *34*, 773–785. [CrossRef] [PubMed]

36. Karpiouk, A.B.; Aglyamov, S.R.; Mallidi, S.; Shah, J.; Scott, W.G.; Rubin, J.M.; Emelianov, S.Y. Combined ultrasound and photoacoustic imaging to detect and stage deep vein thrombosis: Phantom and ex vivo studies. *J. Biomed. Opt.* **2008**, *13*, 054061. [CrossRef] [PubMed]

37. Ranger, B.; Littrup, P.J.; Duric, N.; Chandiwala-Mody, P.; Li, C.P.; Schmidt, S.; Lupinacci, J. Breast ultrasound tomography versus MRI for clinical display of anatomy and tumor rendering: Preliminary results. *Am. J. Roentgenol.* **2012**, *198*, 233–239. [CrossRef] [PubMed]

38. Duric, N.; Littrup, P.; Babkin, A.; Chambers, D.; Azevedo, S.; Kalinin, A.; Pevzner, R.; Tokarev, M.; Holsapple, E.; Rama, O.; et al. Development of ultrasound tomography for breast imaging: Technical assessment. *Med. Phys.* **2005**, *32*, 1375–1386. [CrossRef] [PubMed]

39. Alshahrani, S.; Yan, Y.; Avrutsky, I.; Anastasio, M.; Malyarenko, E.; Duric, N.; Mehrmohammadi, M. Design and development of a full-ring ultrasound and photoacoustic tomography system for breast cancer imaging. In Proceedings of the 2017 IEEE International Ultrasonics Symposium (IUS), Washington, DC, USA, 6–9 September 2017.

40. Alshahrani, S.; Pattyn, A.; Alijabbari, N.; Yan, Y.; Anastasio, M.; Mehrmohammadi, M. The Effectiveness of the Omnidirectional Illumination in Full-Ring Photoacoustic Tomography. In Proceedings of the 2018 IEEE International Ultrasonics Symposium (IUS), Kobe, Japan, 22–25 October 2018.

41. Alshahrani, S.S.; Yan, Y.; Malyarenko, E.; Avrutsky, I.; Anastasio, M.A.; Mehrmohammadi, M. An advanced photoacoustic tomography system based on a ring geometry design. In *Medical Imaging 2018: Ultrasonic Imaging and Tomography*; International Society for Optics and Photonics: Bellingham, WA, USA, 2018; Volume 10580, p. 1058005.

42. Alshahrani, S.S.; Yan, Y.; Alijabbari, N.; Pattyn, A.; Avrutsky, I.; Malyarenko, E.; Poudel, J.; Anastasio, M.; Mehrmohammadi, M. All-reflective ring illumination system for photoacoustic tomography. *J. Biomed. Opt.* **2019**, *24*, 046004.

43. Maggi, L.; Cortela, G.; von Kruger, M.A.; Negreira, C.; de Albuquerque Pereira, W.C. Ultrasonic Attenuation and Speed in phantoms made of PVCP and Evaluation of acoustic and thermal properties of ultrasonic phantoms made of polyvinyl chloride-plastisol (PVCP). In Proceedings of the IWBBIO, Granada Spain, 12–20 March 2013.

44. Li, C.; Sandhu, G.Y.; Boone, M.; Duric, N. Breast imaging using waveform attenuation tomography. In *Medical Imaging 2017: Ultrasonic Imaging and Tomography*; International Society for Optics and Photonics: Bellingham, WA, USA, 2017; Volume 10139, p. 101390A.

45. Xu, M.; Wang, L.V. Universal back-projection algorithm for photoacoustic computed tomography. *Phys. Rev. E* **2005**, *71*, 016706. [CrossRef]

applied
sciences

Article

Self-Gated Respiratory Motion Rejection for Optoacoustic Tomography

Avihai Ron [1,2], Neda Davoudi [3,4], Xosé Luís Deán-Ben [3,5] and Daniel Razansky [1,2,3,5,*,†]

1   Institute for Biological and Medical Imaging, Helmholtz Center Munich, 85764 Neuherberg, Germany
2   Faculty of Medicine, Technical University of Munich, 81765 Munich, Germany
3   Institute for Biomedical Engineering and Department of Information Technology and Electrical Engineering, ETH Zurich, 8093 Zurich, Switzerland
4   Department of Informatics, Technical University of Munich, 85748 Garching, Germany
5   Faculty of Medicine and Institute of Pharmacology and Toxicology, University of Zurich, 8057 Zurich, Switzerland
*   Correspondence: daniel.razansky@uzh.ch; Tel.: +41-44-633-34-29
†   Current Address: Institute for Biomedical Engineering, University of Zurich and ETH Zurich, HIT E42.1, Wolfgang-Pauli-Strasse 27, CH 8093 Zurich, Switzerland.

Received: 29 May 2019; Accepted: 3 July 2019; Published: 6 July 2019

Abstract: Respiratory motion in living organisms is known to result in image blurring and loss of resolution, chiefly due to the lengthy acquisition times of the corresponding image acquisition methods. Optoacoustic tomography can effectively eliminate *in vivo* motion artifacts due to its inherent capacity for collecting image data from the entire imaged region following a single nanoseconds-duration laser pulse. However, multi-frame image analysis is often essential in applications relying on spectroscopic data acquisition or for scanning-based systems. Thereby, efficient methods to correct for image distortions due to motion are imperative. Herein, we demonstrate that efficient motion rejection in optoacoustic tomography can readily be accomplished by frame clustering during image acquisition, thus averting excessive data acquisition and post-processing. The algorithm's efficiency for two- and three-dimensional imaging was validated with experimental whole-body mouse data acquired by spiral volumetric optoacoustic tomography (SVOT) and full-ring cross-sectional imaging scanners.

Keywords: optoacoustic imaging; photoacoustic tomography; respiratory gating; motion artifacts

---

1. Introduction

Motion during signal acquisition is known to result in image blurring and can further hinder proper registration of images acquired by different modalities [1–4]. Respiratory motion compensation in tomographic imaging methods is often based on a gated acquisition assisted by physiological triggers, e.g., an electrocardiogram (ECG) signal. In prospective gating, the data is acquired during a limited time window when minimal, or no motion occurs. Alternately, retrospective gating correlates between the acquired images and physiological triggers during post-processing [5]. More advanced retrospective approaches are based on self-gated methods where the physiological trigger is extracted from the image data itself [6–8]. An alternative solution consists in motion tracking of specific points and subsequent correction with rigid-body transformations [9]. In some parts of the body, such as the thoracic region, non-rigid motion is further produced. Thus, more sophisticated models are generally required to estimate and correct for the effects of respiratory motion [10].

High-frame-rate imaging modalities can avoid motion if sub-pixel displacements are produced during the effective image integration time. Particularly, optoacoustic tomography (OAT) can render 2D and 3D images via excitation of an entire volume with a single laser pulse [11]. This corresponds to an effective integration time in the order of the pulse duration, typically a few nanoseconds. This way, the tissue motion can be "frozen" much more efficiently than most other imaging modalities. OAT has found applicability in biological studies demanding high-frame-rate imaging, such as characterization of cardiac dynamics [12], mapping of neuronal activity [13], monitoring hemodynamic patterns in tumors [14] or visualization of freely-behaving animals [15]. Moreover, real-time imaging has been paramount in the successful translation of OAT to render motion-free images acquired in a handheld mode [16]. While motion correction in OAT might not be relevant for images rendered with a single laser pulse, acquisition of multiple frames is still required in many applications, e.g., for rendering volumetric data from multiple cross-sections or for extending the effective field of view (FOV) of a given imaging system [17]. Multiple frames are also required for multi-spectral optoacoustic tomography (MSOT) applications, where mapping of intrinsic tissue chromophores or extrinsically administered agents is achieved via spectral or temporal unmixing [18–21]. Cardiac and breathing motion could readily be captured by OAT systems running at frame rates of tens of hertz [22,23], and several approaches have been suggested to mitigate motion artefacts in applications involving multi-frame data analysis. For instance, respiratory motion gating was suggested by simultaneously capturing the animal's respiratory waveforms [24]. Motion correction was alternatively performed with 3D rigid-body transformations [25] and with free-form deformation models [26]. Models of body motion have also been suggested for other types of scanning-based systems [27,28]. Additionally, motion suppression could be achieved by reducing the delay between consecutive pulsed light excitations [29,30], which requires dedicated laser systems.

In this work, we demonstrate that motion rejection in OAT can effectively be performed on-the-fly, before image reconstruction. The suggested approach consists in clustering a sequence of OAT frames that employs the raw time-resolved signals without involving computationally and memory extensive post-processing. This represents an important advantage over other known approaches operating in the image domain [31].

2. Materials and Methods

2.1. Pre-Reconstruction Motion Rejection Approach

The algorithm suggested in this work aims at motion rejection in OAT systems based on a multi-frame acquisition of time-resolved pressure signals with transducer arrays. Figure 1a schematically depicts two examples of transducer array configurations for 2D and 3D imaging, which are described in more detail in the following sections. The acquired signals are generally arranged into so-called sinograms, where every sinogram represents a single frame (Figure 1b). At a fixed transducer position, k frames (sinograms) are acquired. These frames consist of matrices with rows representing the m time-samples of each signal and columns corresponding to the n transducer elements (channels) of the array. Step 1 of the algorithm consists in rearranging the k frames of the sequence into columns of a 2D matrix containing m x n rows and k columns, which represent the entire sequence of frames at a fixed transducer position (Figure 1c). In the experiments performed, the number of frames acquired at each array position were chosen to adequately capture a complete breathing cycle. Step 2 of the algorithm consists in calculating the autocorrelation matrix of all pairs of frames (MATLAB (Mathworks Inc, Natick, USA) function 'corrcoef'). An example of the calculated correlation coefficients is displayed in Figure 1d. At a fixed transducer position, time decorrelation is expected to be chiefly affected by respiratory motion. Clustering of frames is subsequently done in Step 3 by applying the second order k-means method to the correlation coefficients matrix (Figure 1e, MATLAB function 'kmeans'). In Step 4, the k frames are then divided into two sets based on predetermined knowledge of the characteristic physiology of the animal under specific anesthesia. As a rule, motion frames are typically fewer than static frames. Notably, when scanning at multiple transducer positions, Steps 1–4 are to be repeated for each transducer position. As an example, Figure 1f displays a comparison of the 3D views of a reconstructed image from a single position of the spherical array, as obtained from the averaged selected-frames and from the averaged rejected-frames.

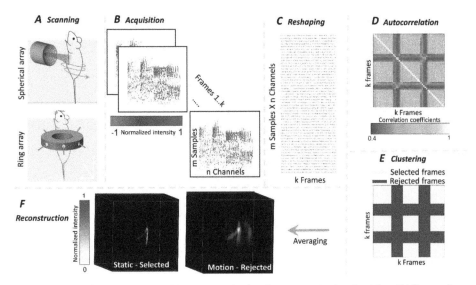

Figure 1. A schematic diagram of the steps involved in the motion rejection algorithm. (**A**) Two- and three-dimensional scanning systems, (top) spiral volumetric optoacoustic tomography (SVOT) based on a spherical array of transducers and (bottom) cross-sectional optoacoustic tomography based on a full-ring array of cylindrically focused transducers. (**B**) Sequence of frames (sinograms) acquired at a single position of the scanner. (**C**) Rearrangement of the data corresponding to the entire sequence into a single matrix. (**D**) Correlation coefficients of the autocorrelation matrix of the columns in (**C**). (**E**) K-means clustering of the correlation coefficients matrix into two groups, namely, selected (static) frames and rejected (motion) frames. (**F**) Volumetric image of a blood vessel reconstructed with data from the selected versus the rejected-frames.

2.2. Spiral Volumetric Optoacoustic Tomography

The spiral volumetric optoacoustic tomography (SVOT) scanner is schematically depicted in Figure 1a (top). A detailed description of the system is available elsewhere [32]. Briefly, a spherical ultrasound array of piezocomposite elements (Imasonics SaS, Voray, France) is mounted on motorized rotating and translating stages and scanned around the animal following a spiral trajectory. The array consists of 256 elements with a central frequency of 4 MHz and −6 dB bandwidth of ~100%, arranged in a hemispherical surface with angular coverage of 90°. The excitation light beam is guided via a fiber bundle (CeramOptec GmbH, Bonn, Germany) through a cylindrical aperture in the center of the sphere. SVOT enables imaging of the entire mouse with a nearly isotropic 3D spatial resolution in the 200 μm range [31]. In the experiments, light excitation was provided with a short-pulsed laser (<10 ns duration pulses with 25 mJ per-pulse energy and up to 100 Hz pulse repetition frequency) based on an optical parametric oscillator (OPO) crystal (Innolas GmbH, Krailling, Germany). The pulse repetition frequency of the laser was set to 25 Hz and the wavelength was maintained at 800 nm, corresponding to the isosbestic point of hemoglobin. The array was scanned for 17 angular positions separated by 15° (total angular coverage in the azimuthal direction of 240°) and for 30 vertical positions separated by 2 mm (total scanning length of 58 mm, a full-body scan requires approximately 10 min). 50 frames were captured for each position of the array, for which all signals were simultaneously digitized at 40 megasamples per second with a custom-made data acquisition system (DAQ, Falkenstein Mikrosysteme GmbH, Taufkirchen, Germany) triggered with the Q-switch output of the laser. The acquired data was eventually transmitted to a PC via Ethernet.

2.3. Cross-Sectional Optoacoustic Tomography with a Ring Array

The system layout is depicted in Figure 1a (bottom) while its detailed description is available in [33]. Briefly, the ultrasound array (Imasonics SaS, Voray, France) consists of an 80 mm diameter ring having 512 ultrasound individual detection elements with 5 MHz central frequency and −6 dB bandwidth of ~80%. Each element is cylindrically focused at a distance of 38 mm to selectively capture signals from the imaged cross-section. In the experiments, light excitation was provided by a short-pulsed (<10 ns duration pulses at a wavelength of 1064 nm with ~100 mJ per-pulse energy and 15 Hz pulse repetition frequency) Nd:YAG laser (Spectra Physics, Santa Clara, California). The laser beam was guided with a fiber bundle (CeramOptec GmbH, Bonn, Germany) having 12 output arms placed around the circumference of the ring transducer with an angular separation of 60° between the arms. Much like the SVOT system, signals detected by all the array elements were simultaneously digitized at 40 megasamples per second using custom-made DAQ (Falkenstein Mikrosysteme GmbH, Taufkirchen, Germany), triggered with the Q-switch output of the laser. The data was transmitted to a PC via Ethernet. In total, 100 frames were recorded with the array positioned at two distinct regions of the mouse.

2.4. Image Reconstruction and Processing

In both scanning systems, the acquired signals were band-pass filtered (cut-off frequencies 0.25–6 MHz for SVOT and 0.5–8 MHz for cross-sectional OAT) and deconvolved with the impulse response of the array elements before reconstruction. For SVOT, tomographic reconstructions of single volumes ($15 \times 15 \times 15$ mm^3) for each scanning position of the spherical array transducer were done using a 3D back-projection-based algorithm [34,35]. Volumetric images reconstructed at every transducer position were stitched together to render images from a larger field of view (whole-body scale). For cross-sectional OAT, the same back-projection algorithm was modified to account for the heterogeneous distribution of the speed of sound in the mouse versus the coupling medium (water) [36]. For this, an initial image was first reconstructed by considering a uniform speed of sound corresponding to the speed of sound in water (determined from the measured water temperature). The animal's surface was then manually segmented, and the reconstruction was fine-tuned by assigning a different speed of sound to the segmented tissue volume in order to optimize image quality. The processing was executed with a self-developed MATLAB code. Universal image quality index (QI) was calculated for the resulting images. QI is an objective image quality index that combines three models: loss of correlation, luminance distortion and contrast distortion—a detailed description and efficient MATLAB implementation was reported in [37].

2.5. Mouse Experiments

All *in-vivo* animal experiments were performed in full compliance with the institutional guidelines of the Helmholtz Center Munich and with approval by the Government District of Upper Bavaria. Hairless NOD.SCID mice (Envigo, Rossdorf, Germany) were anesthetized with isoflurane. For both imaging systems, a custom-made holder was used to vertically fix the mice in a stationary position with fore and hind paws attached to the holder during the experiments. The mice were immersed inside the water tank with the animal head being kept above water. The temperature of the water tank was maintained at 34 °C with a feedback controlled heating stick. A breathing mask with a mouth clamp was used to fix the head in an upright position and to supply anesthesia and oxygen. During measurement, the anesthesia level was kept at ~2% isoflurane.

3. Results

3.1. Spiral Volumetric Optoacoustic Tomography

A whole-body (neck to hind paws) SVOT image, reconstructed using the frames selected with the proposed motion rejection approach, is displayed in Figure 2a. In these experiments, an average of 32% of frames were rejected per transducer position. The effectiveness of the motion rejection approach is demonstrated by analyzing three specific regions (dashed squares in Figure 2a). Specifically, we compare the image combining all the frames, with the one obtained by averaging the selected (static) frames as well as the image obtained by averaging the rejected (motion) frames. Note that the red square partially captures the thoracic region. It can be observed that a small vessel, clearly visible in the image rendered from the selected frames (Figure 2b, red arrow), cannot be resolved in the image reconstructed using all the frames. Furthermore, the former image features a clear motion artifact in the form of a 'double vessel' (red arrow), thus concealing the small vessel. The green square captures the region around the liver. A vertical vessel appears regular and complete in the selected-frames image (red arrow), whereas the same vessel appears disrupted in the all-frames image. Also here, the rejected-frames image discloses an artifact responsible for distorting the all-frames image. Finally, the blue square captures part of the abdomen. Clearly, small vessels are better resolved in the selected-frames image (red arrows). Notably, different structures appear to be blurred (yellow arrows) in the rejected-frames images with respect to the selected-frames images. A comparison between amplitude profiles of structures labeled by dashed yellow lines in Figure 2b further emphasizes the effectiveness of the motion rejection algorithm (Figure 2c) with the signal amplitude typically improved by 10% to 30% in the selected-frames images. Likewise, fine details appear more prominent in the selected-frames images, which is evinced by additional, fine peaks in the amplitudes profiles.

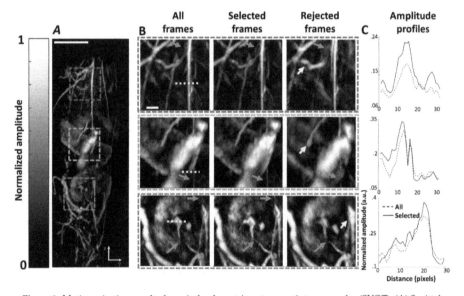

Figure 2. Motion rejection results for spiral volumetric optoacoustic tomography (SVOT). (**A**) Sagittal maximal intensity projection (MIP) of a volumetric image of the mouse reconstructed with the selected-frames (scale bar—1 cm). (**B**) Zoom-in of three regions marked in red, green, blue, respectively, in (**A**). Each image is reconstructed with (left) all-frames, (center) the selected (static) frames (right) the rejected (motion) frames (scale bar—1 mm). Structural differences are marked (yellow and red arrows). (**C**) Amplitude profiles marked in b (yellow dashed lines) for images reconstructed from all the frames (dashed lines) versus selected frames (solid lines).

3.2. Cross-Sectional Optoacoustic Tomography

Effectiveness of the algorithm in cross-sectional OAT was tested by comparing the selected- and rejected-frames images taken from two distinct regions of the animal (Figure 3a). Between 20% and 31% of the frames were rejected in the top and bottom cross-sections, respectively. The rejected-frames images reveal smearing artifacts caused by a breathing motion that are evident across the entire mouse cross-section. Fine structures (red arrows) within the abdominal space appear blurred in the rejected-frames image. Moreover, some superficial structures seem to be artificially 'doubled' (yellow arrows) in the rejected-frames images. Minor differences were observed in the all-frames images (data not shown) with respect to the selected-frames images. Likewise, amplitude profiles from selected structures (dashed yellow line in Figure 3a) are increased by ~10% in the selected-frames images (Figure 3b). The calculated QI clearly reveals distortions at the boundaries of major structures, located mostly superficially (Figure 3c).

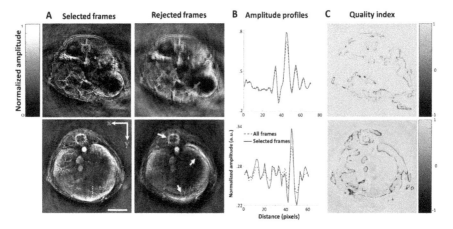

Figure 3. Motion rejection results for cross-sectional imaging with the ring array system. (**A**) Reconstructed transverse slices of a mouse for two different locations rendered by considering the selected (left) versus rejected (right) frames (scale bar—1 cm). Distorted structures are marked (red and yellow arrows). (**B**) Amplitude profiles (of yellow dashed lines in (**A**)) for the images rendered with all (dashed line) versus selected (solid line) frames. (**C**) Distortion-based QI of the difference between the selected- and all-frames images (1 = high similarity; −1 = low similarity).

4. Discussion

The presented results demonstrate that motion rejection in OAT can effectively be accomplished prior to image reconstruction. This represents a significant advantage with respect to previously reported motion rejection approaches based on auto-correlation of a sequence of reconstructed images [31], which are afflicted with excessive memory and post-processing requirements. The suggested method was successfully validated with data acquired by two- and three-dimensional imaging systems. However, motion rejection was more effective in the case of volumetric SVOT scans. In particular, it benefited from both amplitude increase of 10% to 30% and improvement in the visibility of fine details, whereas images from the cross-sectional imaging system yielded a lower amplitude increase (~10%) and minor improvement in the visibility of structures. The reason behind the reduced performance of motion rejection in cross-sectional imaging may be ascribed to the fact that breathing-associated movements are not limited to a single plane, while in-plane motion is mainly detected in the signals. Yet, although the differences between selected- and all-frames cross-sectional images were minor, it was possible to quantify them by utilizing a QI based distortion measures. Notably, such distortion artifacts affect almost exclusively the edges of large structures. In spite of the fact that standard frame averaging in cross-sectional imaging may yield qualitatively comparable results, reliable rejection of 20% to 31% motion-affected frames by the algorithm may turn crucial for quantitative analyses of high resolution data, e.g., involving spectral unmixing of fine structures.

It is also important to take into account that breathing characteristics may differ from one animal to another due to age, health, size, sex or strain. All these factors affect the resilience of the animal to the experimental setup, the feasible depth of anesthesia and the overall duration of the experiment [38]. It was previously reported that mice under 2% isoflurane anesthesia have an average respiratory rate of 44 ± 9 breaths/min [39], where the breathing rhythm is characterized by pauses between breaths longer than the breaths themselves. As a result, the majority of the frames are static, i.e., not affected by motion. Herein, we relied on such prior knowledge of the characteristic respiratory rate and breathing rhythm to establish a rejection criterion for the clustered motion (rejected) frames. Likewise, other criteria independent of these factors may alternatively be implemented.

In conclusion, the developed motion rejection methodology can benefit numerous optoacoustic imaging methods relying on multi-frame image analysis, such as scanning-based tomography or spectroscopic imaging systems like the MSOT. It may also find applicability in handheld clinical imaging [40,41], where motion can hinder accurate signal quantification and interpretation of longitudinal and spectroscopic data.

Author Contributions: Conceptualization, A.R., X.L.D.-B. and D.R.; methodology, A.R.; software, A.R.; validation, A.R. and N.D.; formal analysis, A.R.; investigation, A.R.; resources, D.R.; data curation, A.R. and X.L.D.-B.; writing—original draft preparation, A.R., X.L.D.-B. and D.R.; writing—review and editing, A.R., X.L.D.-B. and D.R.; visualization, A.R.; supervision, D.R.; project administration, D.R.; funding acquisition, D.R.

Funding: This research received no external funding.

Acknowledgments: The authors wish to thank M. Reiss for his support with the measurements and handling of animals.

Conflicts of Interest: The authors declare no conflict of interest.

References

1. Nehmeh, S.A.; Erdi, Y.E. Effect of respiratory gating on quantifying PET images of lung cancer. *J. Nucl. Med.* **2002**, *43*, 876–881.
2. Chi, P.-C.M.; Mawlawi, O. Effects of respiration-averaged computed tomography on positron emission tomography/computed tomography quantification and its potential impact on gross tumor volume delineation. *Int. J. Radiat. Oncol. Biol. Phys.* **2008**, *71*, 890–899. [CrossRef] [PubMed]
3. Liu, C.; Pierce, L.A. II The impact of respiratory motion on tumor quantification and delineation in static PET/CT imaging. *Phys. Med. Biol.* **2009**, *54*, 7345. [CrossRef] [PubMed]
4. Nehrke, K.; Bornert, P. Free-breathing cardiac MR imaging: Study of implications of respiratory motion—Initial results. *Radiology* **2001**, *220*, 810–815. [CrossRef] [PubMed]
5. Heijman, E.; de Graaf, W. Comparison between prospective and retrospective triggering for mouse cardiac MRI. *NMR Biomed.* **2007**, *20*, 439–447. [CrossRef] [PubMed]
6. Zaitsev, M.; Maclaren, J. Motion artifacts in MRI: A complex problem with many partial solutions. *J. Magn. Reson. Imaging* **2015**, *42*, 887–901. [CrossRef] [PubMed]
7. Sureshbabu, W.; Mawlawi, O. PET/CT imaging artifacts. *J. Nucl. Med. Technol.* **2005**, *33*, 156–161.
8. Nehmeh, S.A.; Erdi, Y.E. *Respiratory Motion in Positron Emission Tomography/Computed Tomography: A Review*; Elsevier: Amsterdam, The Netherlands, 2008; pp. 167–176.
9. Maclaren, J.; Herbst, M. Prospective motion correction in brain imaging: A review. *Magn. Reson. Med.* **2013**, *69*, 621–636. [CrossRef]
10. McClelland, J.R.; Hawkes, D.J. Respiratory motion models: A review. *Med. Image Anal.* **2013**, *17*, 19–42. [CrossRef]
11. Deán-Ben, X.; Gottschalk, S. Advanced optoacoustic methods for multiscale imaging of in vivo dynamics. *Chem. Soc. Rev.* **2017**, *46*, 2158–2198. [CrossRef]
12. Lin, H.-C.A.; Deán-Ben, X.L. Characterization of Cardiac Dynamics in an Acute Myocardial Infarction Model by Four-Dimensional Optoacoustic and Magnetic Resonance Imaging. *Theranostics* **2017**, *7*, 4470. [CrossRef] [PubMed]
13. Gottschalk, S.; Degtyaruk, O. Rapid volumetric optoacoustic imaging of neural dynamics across the mouse brain. *Nat. Biomed. Eng.* **2019**, *3*, 392–401. [CrossRef] [PubMed]
14. Ron, A.; Deán-Ben, X.L. Volumetric optoacoustic imaging unveils high-resolution patterns of acute and cyclic hypoxia in a murine model of breast cancer. *Cancer Res.* **2019**. [CrossRef] [PubMed]

15. Özbek, A.; Deán-Ben, X.L. Optoacoustic imaging at kilohertz volumetric frame rates. *Optica* **2018**, *5*, 857–863. [CrossRef]
16. Neuschmelting, V.; Burton, N.C. Performance of a multispectral optoacoustic tomography (MSOT) system equipped with 2D vs. 3D handheld probes for potential clinical translation. *Photoacoustics* **2016**, *4*, 1–10. [CrossRef] [PubMed]
17. Deán-Ben, X.L.; López-Schier, H. Optoacoustic micro-tomography at 100 volumes per second. *Sci. Rep.* **2017**, *7*, 6850. [CrossRef] [PubMed]
18. Ron, A.; Deán-Ben, X.L. Characterization of Brown Adipose Tissue in a Diabetic Mouse Model with Spiral Volumetric Optoacoustic Tomography. *Mol. Imaging Biol.* **2018**. [CrossRef] [PubMed]
19. Taruttis, A.; Ntziachristos, V. Advances in real-time multispectral optoacoustic imaging and its applications. *Nat. Photonics* **2015**, *9*, 219. [CrossRef]
20. Deán-Ben, X.L.; Stiel, A.C. Light fluence normalization in turbid tissues via temporally unmixed multispectral optoacoustic tomography. *Opt. Lett.* **2015**, *40*, 4691–4694. [CrossRef]
21. Yao, J.; Kaberniuk, A.A. Multiscale photoacoustic tomography using reversibly switchable bacterial phytochrome as a near-infrared photochromic probe. *Nat. Methods* **2016**, *13*, 67. [CrossRef]
22. Wang, L.; Maslov, K.I. Video-rate functional photoacoustic microscopy at depths. *J. Biomed. Opt.* **2012**, *17*, 106007. [CrossRef] [PubMed]
23. Taruttis, A.; Claussen, J. Motion clustering for deblurring multispectral optoacoustic tomography images of the mouse heart. *J. Biomed. Opt.* **2012**, *17*, 016009. [CrossRef] [PubMed]
24. Xia, J.; Chen, W. Retrospective respiration-gated whole-body photoacoustic computed tomography of mice. *J. Biomed. Opt.* **2014**, *19*, 016003. [CrossRef] [PubMed]
25. Gottschalk, S.; Fehm, T.F. Correlation between volumetric oxygenation responses and electrophysiology identifies deep thalamocortical activity during epileptic seizures. *Neurophotonics* **2016**, *4*, 011007. [CrossRef]
26. Toi, M.; Asao, Y. Visualization of tumor-related blood vessels in human breast by photoacoustic imaging system with a hemispherical detector array. *Sci. Rep.* **2017**, *7*, 41970. [CrossRef] [PubMed]
27. Schwarz, M.; Garzorz-Stark, N. Motion correction in optoacoustic mesoscopy. *Sci. Rep.* **2017**, *7*, 10386. [CrossRef] [PubMed]
28. Chung, J.; Nguyen, L. Motion estimation and correction in photoacoustic tomographic reconstruction. *SIAM J. Imaging Sci.* **2017**, *10*, 216–242. [CrossRef]
29. Deán-Ben, X.L.; Bay, E. Functional optoacoustic imaging of moving objects using microsecond-delay acquisition of multispectral three-dimensional tomographic data. *Sci. Rep.* **2014**, *4*, 5878. [CrossRef]
30. Märk, J.; Wagener, A. Photoacoustic pump-probe tomography of fluorophores in vivo using interleaved image acquisition for motion suppression. *Sci. Rep.* **2017**, *7*, 40496. [CrossRef]
31. Fehm, T.F.; Deán-Ben, X.L. In vivo whole-body optoacoustic scanner with real-time volumetric imaging capacity. *Optica* **2016**, *3*, 1153–1159. [CrossRef]
32. Deán-Ben, X.L.; Fehm, T.F. Spiral volumetric optoacoustic tomography visualizes multi-scale dynamics in mice. *Light Sci. Appl.* **2017**, *6*, e16247. [CrossRef] [PubMed]
33. Merčep, E.; Herraiz, J.L. Transmission–reflection optoacoustic ultrasound (TROPUS) computed tomography of small animals. *Light Sci. Appl.* **2019**, *8*, 18. [CrossRef] [PubMed]
34. Xu, M.; Wang, L.V. Universal back-projection algorithm for photoacoustic computed tomography. *Phys. Rev. E* **2005**, *71*, 016706. [CrossRef] [PubMed]
35. Ozbek, A.; Deán-Ben, X. *Realtime Parallel Back-Projection Algorithm for Three-Dimensional Optoacoustic Imaging Devices*; Optical Society of America: Washington, DC, USA, 2013; p. 88000I.
36. Deán-Ben, X.L.; Özbek, A. Accounting for speed of sound variations in volumetric hand-held optoacoustic imaging. *Front. Optoelectron.* **2017**, *10*, 280–286. [CrossRef]

37. Wang, Z.; Bovik, A.C. A universal image quality index. *IEEE Signal Process. Lett.* **2002**, *9*, 81–84. [CrossRef]
38. Gargiulo, S.; Greco, A. Mice anesthesia, analgesia, and care, Part II: Anesthetic considerations in preclinical imaging studies. *ILAR J.* **2012**, *53*, E70–E81. [CrossRef]
39. Kober, F.; Iltis, I. Cine-MRI assessment of cardiac function in mice anesthetized with ketamine/xylazine and isoflurane. *Magn. Reson. Mater. Phys. Biol. Med.* **2004**, *17*, 157–161. [CrossRef]
40. Diot, G.; Metz, S. Multi-Spectral Optoacoustic Tomography (MSOT) of human breast cancer. *Clin. Cancer Res.* **2017**, *23*, 6912–6922. [CrossRef]
41. Reber, J.; Willershäuser, M. Non-invasive Measurement of Brown Fat Metabolism Based on Optoacoustic Imaging of Hemoglobin Gradients. *Cell Metab.* **2018**, *27*, 689–701. [CrossRef]

Article

Accelerated Correction of Reflection Artifacts by Deep Neural Networks in Photo-Acoustic Tomography

Hongming Shan [1], Ge Wang [1] and Yang Yang [2,*]

[1] Biomedical Imaging Center, Department of Biomedical Engineering, Rensselaer Polytechnic Institute, Troy, NY 12180, USA

[2] Department of Computational Mathematics, Science and Engineering, Michigan State University, East Lansing, MI 48824, USA

* Correspondence: yangy5@msu.edu; Tel.: +1-517-432-0620

Received: 20 May 2019; Accepted: 18 June 2019; Published: 28 June 2019

Abstract: Photo-Acoustic Tomography (PAT) is an emerging non-invasive hybrid modality driven by a constant yearning for superior imaging performance. The image quality, however, hinges on the acoustic reflection, which may compromise the diagnostic performance. To address this challenge, we propose to incorporate a deep neural network into conventional iterative algorithms to accelerate and improve the correction of reflection artifacts. Based on the simulated PAT dataset from computed tomography (CT) scans, this network-accelerated reconstruction approach is shown to outperform two state-of-the-art iterative algorithms in terms of the peak signal-to-noise ratio (PSNR) and the structural similarity (SSIM) in the presence of noise. The proposed network also demonstrates considerably higher computational efficiency than conventional iterative algorithms, which are time-consuming and cumbersome.

Keywords: photo-acoustic tomography; reflection artifacts; deep learning; convolutional neural network; time reversal; Landweber algorithm; U-net

1. Introduction

Photo-Acoustic Tomography (PAT) is an emerging non-invasive modality that has manifested an enormous prospect for some clinical practices [1]. In PAT, the tissue is illuminated with near-infrared light of wavelength 650–900 nm. The absorbed optical energy is transformed into acoustic energy through the photo-acoustic effect, and the generated ultrasound is measured by transducer arrays outside the tissue to retrieve the optical properties of the tissue. The coupling mechanism of optical and ultrasound waves gives multiple advantages over conventional individual imaging modalities. As the acoustic waves experience much less scattering in tissue compared to optical waves, PAT can generate high-resolution images in the presence of strong optical scattering to break the optical diffusion limit [2]. The image can reach submillimeter spatial resolution while preserving intrinsic optical contrast [3].

Typical photo-acoustic signal generation comprises three steps: (1) the tissue absorbs light; (2) the absorbed optical energy causes a temperature rise; (3) thermo-elastic expansion occurs and generates ultrasound. The image formation in PAT is to recover the distribution of the deposited energy, known as the local optical fluence, from the ultrasound signals that are recorded by the sensors deployed around the tissue. As the initial ultrasound pressure is approximately proportional to the optical fluence, it is sufficient to reconstruct the initial pressure from the recorded ultrasound signals.

The quality of PAT images relies on multiple factors, and one of them is the acoustic reflection. The reflection can be caused by either hyperechoic structures or special setting for making a measurement such as in reverberant-field PAT [4]. Conventional PAT reconstruction algorithms, such as the universal back-projection formula [5] and time-reversal-based reconstruction [6–10], are based on the spherical mean radon transform on canonical geometries and not designed to take the acoustic reflections into account. As a result, the reflected signals are projected along with the real signal back into the image domain, resulting in artifacts that are indistinguishable from the real biological structures. Clinical practitioners who rely on the misleading artifacts to make judgment could come to erroneous conclusions. It is, therefore, essential and of practical significance to design new methods to eliminate the impact of such acoustic reflection in PAT image reconstruction.

Some iterative algorithms were developed to correct the effect of the acoustic reflection. Examples include Fourier analysis [11], averaged time reversal [12], dissipative time reversal [13], and adjoint method [14]. Nonetheless, each of them has its limitation: the Fourier approach applies only to regular domains; averaged/dissipative time reversals, and Landweber iterations can generate high-quality images but time-consuming.

The purpose of this paper is to accelerate and improve the iterative algorithms by incorporating a deep neural network (DNN) structure to reduce the reflection artifacts efficiently and effectively. The entire network contains three parts—feature extraction, artifacts reduction, and reconstruction. A modified version of the U-net with large convolutional filter sizes is used as the backbone of the artifact reduction part to capture the global features by increasing the size of the receptive field. The network is applied to take the place of iterations to accelerate the correction of reflection artifacts. The learning process can simultaneously reduce the image noise as well.

With the rapid increment of computational power, DNNs and deep learning techniques have received considerable attention in recent years for tomographic image reconstruction [15–17]. In particular, they have achieved the state-of-the-art performance in PAT image reconstruction in the scenarios of sparse data [18–20], limited view [21–23], artifacts removal [24–26], as well as other applications [27,28]. It is worth pointing out that the reference [25] considers another type of reflection artifact with different focuses and approaches. The reflection in [25] is caused by point-like targets inside the tissue, while ours by planar reflectors outside the tissue. The network in [25] is mostly based on convolutional neural networks (CNNs) while ours on the U-net structure. The two networks target different problems with associated unique networks and therefore are of independent values. We remark that besides in PAT, DNNs have been successfully applied to other imaging modalities as well, see [29–35] for its applications in CT.

2. Data Generation

This section presents the forward model of PAT and the reflection artifacts induced by the acoustic reflection of signals.

2.1. The Forward Model

We describe the forward model, which is used to generate ultrasound signals for training purpose. Sound hard reflectors, which can bounce the incident ultrasound back with the opposite speed without absorbing any energy, are placed around the tissue to simulate the acoustic reflections. As such reflectors do not dissipate energy, the resultant reflection artifacts are the strongest, hence can be used as a benchmark to test the artifact-correction capability of different algorithms.

Denote by Ω the biological tissue, which is illuminated by rapid pulses of laser, and afterwards, emits the ultrasound. The boundary of the tissue is represented using the conventional notation $\partial\Omega$. The forward ultrasound propagation model, in the presence of sound hard reflectors, reads in 2D as follows:

$$\begin{cases} \left(\partial_t^2 - c^2(r)\Delta\right)p &= 0 & \text{in } (0,T) \times \Omega, \\ p|_{t=0} &= p_0(r), & \\ \partial_t p|_{t=0} &= 0, & \\ \partial_\nu p|_{[0,T] \times \partial\Omega} &= 0, & \end{cases} \tag{1}$$

where $\Delta = \frac{\partial^2}{\partial x^2} + \frac{\partial^2}{\partial y^2}$ is the Laplace operator, T is the stoppage time, $p_0(r)$ is the initial ultrasound pressure, $p(t,r)$ is the ultrasound pressure of the spatial location r at time t, $c(r)$ is the wave speed, and ∂_ν is the normal boundary derivative. The main difference of this model with the conventional photo-acoustic models (see, for instance [2,36]) is the boundary condition $\partial_\nu p|_{[0,T] \times \partial\Omega} = 0$, which is the standard mathematical model to describe sound hard reflectors. Please note that the Neumann boundary condition corresponds to the sample/water interface, and the Dirichlet boundary condition, which corresponds to the sample/air interface, can be handled similarly. We remark that modeling PAT using the Neumann and Dirichlet boundary conditions was reported in the literature [37,38]. The ultrasound is recorded at the boundary of Ω using 508 sensors deployed evenly around the tissue. This measurement amounts to the temporal boundary value of p, i.e., $p|_{[0,T] \times \partial\Omega}$. The image formation in PAT aims to recover the distribution of the deposited energy that is the local optical fluence. As the initial ultrasound pressure $p_0(r)$ is proportional to the optical fluence, it remains to reconstruct the initial ultrasound pressure $p_0(r)$ from the recorded ultrasound signals.

We simulate numerous ultrasound signals to train the neural network. This is accomplished by solving the forward model (1) using the second order finite difference time domain (FDTD) method with the central difference formula. The simulation is implemented using the MATLAB code written by the authors. The computational domain is a 128×128 grid with uniform spacing. The time step is chosen according to the spatial step to fulfill the Courant–Friedrichs–Lewy (CFL) condition. The sound speed $c(r)$ is either a constant or a spatially varying function, with the former modeling the propagation in homogeneous media and the latter in heterogeneous media. The stoppage time is $T = 4$, which is chosen in such a way that the ultrasound generated from any interior source can reach the boundary.

2.2. Reflection Artifacts

Applying the conventional PAT reconstruction directly to the ultrasound signals generated by the forward model (1) could result in artifacts. This is because the conventional reconstruction methods do not take into consideration of the acoustic reflections. We provide a numerical example in this section to illustrate the effect of the acoustic reflections.

We choose the Shepp–Logan phantom (see Figure 1a) as the initial pressure $p_0(r)$ to produce ultrasound signals based on the forward model (1) with $c(r) = 1$. The recorded acoustic signal is shown in Figure 1b. To demonstrate the reflection artifacts by conventional algorithms, we adopt one of the conventional time reversal (TR) methods proposed in [8]. The sound speed in the TR is again $c(r) = 1$. Then the TR method is mathematically equivalent to the 2D universal back-projection formula. The stoppage time is again $T = 4$. The reconstructed image is illustrated in Figure 1c. It is clear that direct application of the TR method leads to numerous artifacts in the reconstructed image, especially near the boundary of the ellipses in the Shepp–Logan phantom. The generation of these artifacts can actually be well understood from the mathematical point of view; see the analysis in [12] for details.

The failure of the conventional TR method indicates that some procedures must be introduced to correct the reflection artifacts in the presence of reflections. There have been some iterative correction algorithms, for instance [10–14,39] as well as methods based on deep learning [25]. In the next section, we propose a novel neural network to remove the reflection artifacts. Its efficacy is compared with two of the popular iterative correction algorithms as well.

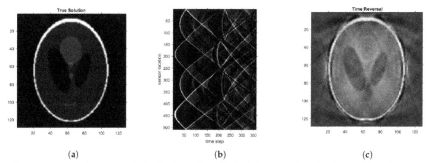

Figure 1. Numerical example to illustrate the effect of the acoustic reflections. (**a**): a 128 × 128 Shepp–Logan phantom as the initial pressure. (**b**): the recorded acoustic signal in the presence of sound hard reflectors. (**c**): reconstruction by conventional TR.

3. Artifacts Correction by Deep Learning

This section presents the proposed neural network for correction of reflection artifacts in PAT imaging. We will make use of the networks to accelerate some conventional iterative algorithms in the following way. Instead of running the iterations, we train the network to learn the map between the reconstructed image by the first iteration and the ground truth. The reconstruction procedure then consists of two steps, given the measured ultrasound signals. The first step is to apply an iterative algorithm to the signals to get the output after the first iteration. The second step is to feed the output into the neural network to obtain the reconstructed image. These two steps preserve the predictability of the model-based iterative algorithm, while at the same time replace the number of iterations by the neural network to achieve improved computational efficiency. The proposed method can be referred to as the deep learning (DL) algorithm or DL reconstruction for short.

3.1. Reflection Artifacts Correction Model

Assume that I_{IS} is the simulated initial source without any artifacts and I_{RA} is a PAT image including reflection artifacts, the relationship between them can be expressed as follows:

$$I_{\mathrm{RA}} = \mathcal{F}(I_{\mathrm{IS}}), \tag{2}$$

where function $\mathcal{F}$ denotes the acoustic-reflection-induced process due to the hyperechoic structures or the special setting of making measurement such as in reverberant-field PAT. The reflection artifacts correction model is to seek an approximate inverse, $\mathcal{G} \approx \mathcal{F}^{-1}$, to reduce the reflection artifacts from I_{RA}, i.e.,

$$I_{\mathrm{est}} = \mathcal{G}(I_{\mathrm{RA}}) \approx I_{\mathrm{IS}}. \tag{3}$$

Next, we introduce the proposed network to learn the approximate inverse function $\mathcal{G}$.

3.2. The Proposed Network

The proposed network architecture for reflection artifacts correction is shown in Figure 2, which has following three parts:

1. **Feature extraction** is to extract the feature representation from the input image I_{RA}. This part contains 4 convolutional layers, each of them has 32 convolutional filters of size 3 × 3 with a stride of 1. Zero padding is not used for these four layers.
2. **Artifacts reduction** is to correct the artifacts and remove the noise from the feature-maps obtained above. We use a modified version of U-net coupling with a residual skip connection for this purpose. This part has 6 convolutional layers and 6 deconvolutional layers. Instead of down-sampling or up-sampling operation in original U-net [40], we use the convolution

with a stride of 2 to decrease the size of feature-maps at the 2nd, 4th and 5th convolutional layers, and the deconvolution with a stride of 2 to increase the size of feature-maps at the 1st, 3rd, and 5th deconvolutional layers. The stride of 1 is used at remaining layers. All layers have 32 (de)convolutional filters of size 5 × 5 to capture the global structures. In addition to this, three conveying links [29,40] copy the former feature-maps and reuse them as the input to a later layer that has the same size of feature-maps via a concatenation operation along channel dimension, which preserve high-resolution features. At last, a residual skip connection [41] enables the above U-net to infer the artifacts and noise from the input feature-maps. The summation between the input feature-maps and the outputs of the U-net are the corrected feature-maps, which is followed by an activation function and then serves as the input to the reconstruction part. Note that zero padding is used throughout this part.

3. **Reconstruction** is to recover the final output from the corrected feature-maps. This part has 4 deconvolutional layers, each of them has 32 filters of size 3 × 3 with a stride of 1 except for the last layer that has only 1 filter. Zero padding is not used for these four layers.

The rectified linear unit (ReLU) is used throughout this network [42], which is defined as $f(x) = x^+ = \max(0, x)$. The input to the network I_{RA} is the output of the first iteration of averaged time reversal (ATR), which is denoted as ATR⋆.

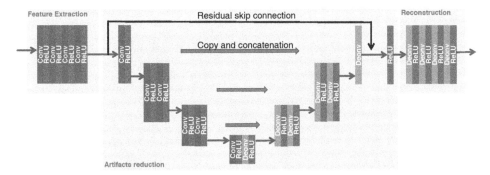

Figure 2. The proposed network structure for reflection artifacts correction. It comprises three parts—feature extraction, artifacts reduction, and reconstruction. In particular, we use a modified version of U-net as the backbone of artifacts reduction part.

3.3. Loss Function

The parameters of the network need to be optimized by minimizing an appropriate loss function. The loss function we choose is the combination of the mean-squared error (MSE) and structural similarity (SSIM) [43].

The MSE between the output of the network, I_{est}, and the initial reference source, I_{IS}, is formally defined as

$$\text{MSE}(I_{est}, I_{IS}) = \frac{1}{w \times h} \left\| I_{est} - I_{IS} \right\|_F^2, \tag{4}$$

where w and h are the width and height of the image, respectively, and $\|\cdot\|$ refers to Frobenius norm. Although MSE is the most straightforward loss function to optimize the network, the resultant images are usually over-smoothing and lose some details.

In contrast to MSE, the SSIM can measure the similarity between two images in terms of their structures and textures. SSIM index is calculated on various windows of an image. The measure between the window x over I_{est} and the window y over I_{IS}, based on a common size $k \times k$, is defined as

$$\text{SSIM}(x, y) = \frac{2\mu_x \mu_y + c_1}{\mu_x^2 + \mu_y^2 + c_1} \frac{2\sigma_{xy} + c_2}{\sigma_x^2 + \sigma_y^2 + c_2}, \tag{5}$$

where μ_x and μ_y are the averages of x and y respectively, σ_x^2 and σ_y^2 are the variances of x and y respectively, and σ_{xy} is the covariance of x and y. Also, $c_1 = 0.01^2$ and $c_2 = 0.03^2$ are two constants, which are used to stabilize the division with a weak denominator. The windows size k is 11, as suggested. The SSIM between two images I_{est} and I_{IS}, $\text{SSIM}(I_{est}, I_{IS})$, refers to the average of the SSIM index over all windows.

The loss function is defined as

$$\min_{\theta_G} \mathbb{E}_{(I_{est}, I_{IS})} \sqrt{1 + \text{MSE}(I_{est}, I_{IS})} \times \left[1 - \text{SSIM}(I_{est}, I_{IS})\right], \tag{6}$$

where θ_G denotes the parameters of the network G. This loss function not only reduces the noise in terms of MSE but also preserve the image structures as measured by SSIM [44]. Various algorithms can solve this minimization problem. In this work, we adopt the Adam algorithm to update the parameters [45]. The gradients of the parameters are computed using a back-propagation algorithm [46].

4. Results

4.1. Experimental Setup

4.1.1. Simulated Dataset

We used 5 cadaver CT scans from Massachusetts General Hospital [47] for simulating PAT dataset. These 5 cadavers were scanned on a GE Discovery 750 HD scanner under 120 kVp x-ray spectra. Noise index (NI) was used by GE to define the image quality, which is approximately equal to the standard deviation of the CT number in the central region of the image of a uniform phantom [48]. In this study, we used a noise index of 10, which represents a normal-dose scanning. Then, the images were reconstructed by the GE commercial iterative reconstruction algorithm, called adaptive statistical iterative reconstruction (ASIR-50%). To reduce the computational cost for simulating the data and training the network, we extracted image patches of size 128×128 from CT scans. More specifically, 64,000 image patches were randomly selected from 3 scans for training purpose. Then, 6400 image patches were randomly selected from 1 scan for validation. Finally, 200 image patches were randomly selected from 1 scan for testing the trained model. Please note that the patients for training, validation, and testing sets were randomly selected from 5 patients CT scans without replacement. Since the Hounsfield unit (HU) used in CT imaging ranges from -1000 to $\sim +2000$, we are interested in the complex tissue structure within $[-160, 240]$ HU window for our PAT simulation. Thus, with this selected HU window, image patches were first normalized into $[0, 1]$ serving as the initial source for PAT imaging, and the output of the first iteration of ATR serves as the input to the network.

4.1.2. Baseline Method

We compare the proposed network-based reconstruction with two state-of-the-art iterative algorithms in correcting reflection artifacts. These iterative algorithms, known as the ATR and adjoint method respectively, are designed to remove the reflection artifacts from the mathematical point of view. The reasons for choosing these two methods are three-fold. Firstly, in contrast to the universal back-projection formula [5] which is mathematically valid only in homogeneous media, these methods can be applied to general heterogeneous media. The universal back-project formula is a special case of the TR method if one sets the sound speed to be constant and applies the Kirchhoff solution formula for the 3D acoustic wave equation. Secondly, the TR and adjoint methods are applicable with arbitrary closed surfaces of sensor arrays. In a typical TR or adjoint process, the measured ultrasound signals are re-transmitted in a temporally reversed order back to the tissue. This can be numerically simulated by solving the wave propagation model backwards to the initial moment while using the measured ultrasound signal as the boundary condition, regardless of the geometry of sensor arrays. Thirdly, TR and adjoint methods are popular in both research and application, for instance [8,9,49,50] for various TR methods and [51–53] for various adjoint methods. Their implementation has also been included in open source packages such as the k-wave MATLAB package [54]. A sketchy introduction to these iterative algorithms is included in the Appendix A for the convenience of the readers.

4.1.3. Parameter Setting

For our proposed network, the initial learning rate λ was set as 1.0×10^{-3} and was adjusted by every epoch, namely $\lambda_t = \frac{\lambda}{\sqrt{t}}$ at the t-th epoch. For the Adam optimization, the coefficients used for computing running averages of gradients and its square were set as 0.9 and 0.999, respectively. The network was implemented with PyTorch DL library [55] and trained within 60 epochs using four NVIDIA GeForce GTX 1080 Ti GPUs. The batch size was set as 512 during the training.

For these two iterative algorithms, the number of iterations for ATR and Landweber is empirically set to be 10. The regularization parameter in Landweber is empirically chosen as 0.03.

4.1.4. Evaluation Metrics

For the evaluation of image quality, we used the peak signal-to-noise ratio (PSNR) and SSIM in our experiments. The SSIM has been defined in (5) and PSNR is defined as follows:

$$\text{PSNR}(I_{\text{est}}, I_{\text{IS}}) = 10 \log_{10} \left(\frac{R^2}{\text{MSE}(I_{\text{est}}, I_{\text{IS}})} \right), \tag{7}$$

where R is the maximum possible pixel value of the images, which is 1 in this study as the images are normalized into $[0, 1]$.

4.2. Results

In this part, we demonstrate the performance of the proposed method in correcting reflection artifacts.

4.2.1. Homogeneous Media

Soft biological tissue is made up mostly of water and therefore can be viewed approximately as a homogeneous medium. The sound speed in water is 1480 m/s, which can be taken as the speed in tissue. The specific value of the constant is irrelevant to the performance of the algorithms, as the value can always be adjusted by choosing different units. What matters is that the speed must be uniform everywhere. Therefore, we simply take the speed as 1 in our numerical test, i.e., we set $c(r) = 1$. Ultrasound signals are generated using the forward model (1). The signals are then exploited in TR, ATR, Landweber iteration, and DL algorithms, respectively, to reconstruct the original image. The input of the neural network is the output of the first iteration of ATR, which is denoted as ATR*.

Based on an independent testing set of 200 CT image patches, we investigated these reconstruction algorithms with 0%, 10%, 20%, 30%, 40% noise added to the ultrasound signals respectively, and reported the box plots of the image quality that was evaluated by PSNR and SSIM in reference to the ground-truth simulated initial source in Figure 3. The experimental results show the DL algorithm has at least two advantages over the others. On the one hand, it is relatively more stable as the noise level increases. Its PSNR outperforms the others at all noise levels, with the only exception of the ideal zero-noise case where ATR provides the best reconstruction. On the other hand, the DL algorithm is time-saving. Although training the neural network takes large amount of time, its output is almost immediate once the training process is accomplished. In contrast, the ATR and Landweber iteration takes several minutes to achieve an image of satisfactory quality.

A few reconstructed images randomly selected from the testing set are displayed in Figure 4. Each column corresponds to the reconstruction using the algorithm labeled at the bottom. The ATR* column consists of images from the first iteration of ATR, which are the inputs to the neural network. The IS column consists of the ground-truth initial source images used to generate the ultrasound signals by the forward model. Among all the algorithms, TR gives the worst result, which can be expected as it does not resolve reflected artifacts. ATR tends to lose detailed information at a high noise level (see the last row). Landweber performs better in resolving noise as it has regularization effect, which helps remove the high-frequency content of the image. The DL algorithm exhibits superior reconstruction in general.

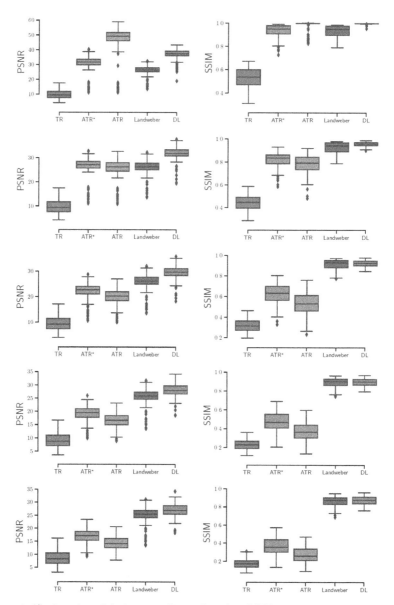

Figure 3. The box plots of the image quality evaluated by PSNR and SSIM on the testing set in a homogeneous medium. From top row to bottom row: noise increases from 0 to 0.4. Diamonds (◇) indicate outliers.

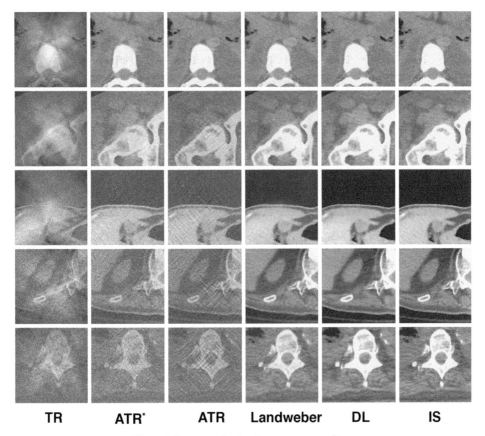

$$\text{TR} \qquad \text{ATR}^* \qquad \text{ATR} \quad \text{Landweber} \quad \text{DL} \qquad \text{IS}$$

Figure 4. Reconstruction in a homogeneous medium.

4.2.2. Heterogeneous Media

Heterogeneity occurs when the object to be imaged has a complex composition. An example is the transcranial PAT where the sound speed in the skull is 3200 m/s in contrast to 1480 m/s in the water. The speed has jump singularities at the interface of the two constituents. Such singularities can be mingled with those from the initial pressure and appear in the reconstructed image as additional artifacts, casting more challenges for the reconstruction of the initial pressure. We implemented the algorithms in the domain $[-1,1] \times [-1,1]$ and chose the distribution of the sound speed as

$$c = 1 - 0.2\sin(2\pi x) + 0.15\cos(\pi y) + \chi_{(x-0.5)^2 + (y-0.5)^2 < 0.01}, \tag{8}$$

where $\chi_{(x-0.5)^2 + (y-0.5)^2 < 0.01}$ is a function that equals 1 on the disk $\{(x,y) : (x-0.5)^2 + (y-0.5)^2 < 0.01\}$ and 0 otherwise. The reason for such choice of c is as follows. The constant 1 models the speed in soft tissue. The smooth term $-0.2\sin(2\pi x) + 0.15\cos(\pi y)$ is added to mimic the slight variation of the sound speed in distinct types of tissue. The non-smooth term $\chi_{(x-0.5)^2 + (y-0.5)^2 < 0.01}$ captures the jumps between different materials such as soft tissue and bones; see Figure 5 for the distribution of the sound speed c.

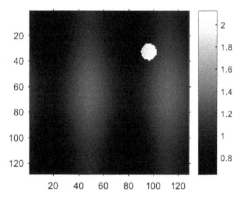

Figure 5. Distribution of the sound speed.

With the variable sound speed, we computed the PSNR and SSIM of the reconstructed images with 0%, 10%, 20%, 30%, 40% noise added to the ultrasound signals respectively, as is shown in Figure 6. The DL algorithm still demonstrates the optimal overall performance. Besides being more stable and time-saving, the DL algorithm also yields the least outliers. Here an outlier is a number in the dataset that is less than $Q_1 - 1.5 \times (Q_3 - Q_1)$ or greater than $Q_3 + 1.5 \times (Q_3 - Q_1)$, where Q_1 is the lower quartile, and Q_3 is the upper quartile.

Some reconstructed images are randomly selected from the testing set again to illustrate the difference of the algorithms; see Figure 7. ATR still suffers from the high noise level, but can resolve the jumps in the speed. Landweber iteration, however, introduces additional artifacts on the top right of the reconstructed image where the sound speed jumps as shown in Figure 5. This is partly due to the limited number of iterations, and it is observed that artifact becomes weaker if the number of iterations is increased. The DL reconstruction resolves simultaneously the high noise and jumps in the speed. It remains visually the closest to the true initial source.

5. Discussions

Our experimental results empirically demonstrated that DL reconstruction in certain situations is superior to the conventional iterative reconstructions, especially when it deals with signals compromised by strong noise. The test on the sound reflectors suggests a great potential of DL-based methods for removing reflection artifacts in the PAT image formation.

However, there are some limitations in this study. First, some of the parameters in these iterative algorithms, such as the number of iterations and the value of the regularization parameter, are not specifically optimized. Varying these parameters may result in somewhat improved performance of the iterative algorithms. However, finding the optimal values of such parameters are highly empirical, and there is no universal approach in general. Second, we empirically used the combination of MSE and SSIM as the loss function to optimize the network, and evaluated the image quality by PSNR and SSIM. The choice of the loss function and image metrics may not be the optimal to capture the visual quality for PAT imaging. Third, we only studied the reflection artifacts under different noise levels and a simple variable sound model, other more complicated conditions such as limited views and more complicated variable sounds can be surely addressed by extending the proposed network.

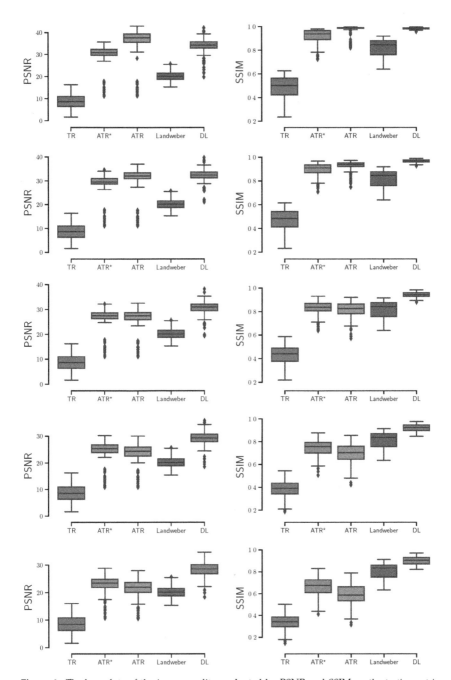

Figure 6. The box plots of the image quality evaluated by PSNR and SSIM on the testing set in a heterogeneous medium. From top row to bottom row: noise increases from 0 to 0.4. Diamonds (◇) indicate outliers.

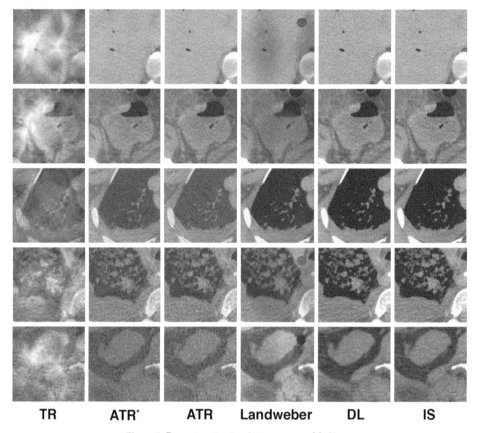

TR   ATR*   ATR   Landweber   DL   IS

Figure 7. Reconstruction in a heterogeneous Medium.

6. Conclusions

In this article, we have proposed a novel DNN to remove the reflection artifacts in reconstructed PAT images under different noise levels and different media. By directly comparing the proposed network to popular iterative reconstruction algorithms with simulated PAT data from CT scans, the results have showed that the proposed network is able to reconstruct superior images over the conventional iterative reconstructions in typical scenarios in terms of computational efficiency and noise reduction.

The results can be further strengthened in several aspects. One practical and significant question is how to make the network robust to potential malignant attacks. It is well known that DL models are vulnerable to adversarial examples. A more stable training procedure is thus critical to the clinical application of DL methods. Next, the effectiveness and efficiency of the network can be further improved for constrained resources and cloud-end processing. Some other factors, such as limited view, acoustic attenuation, and fluctuation of sound speed, can greatly impact the quality of PAT images. It would be interesting to extend the DL approach to these situations as well.

Author Contributions: H.S. and Y.Y. initiated the project and designed the experiments. H.S. performed machine learning research. Y.Y. performed iterative reconstruction research. H.S. and Y.Y. wrote the paper, and G.W. participated in the discussions and edited the paper.

Funding: Y.Y. was partly supported by the NSF grant DMS-1715178, the Simons travel grant, and the startup fund from the Michigan State University.

Acknowledgments: The authors thank NVIDIA Corporation for the donation of GPUs used for this research. The authors would also like to express their gratitude to the anonymous reviewers for the valuable suggestions and comments which helped considerably improve the exposition of the paper.

Conflicts of Interest: H.S. and G.W. have received unrelated industrial research grants from General Electric and Hologic Inc.

Abbreviations

The following abbreviations are used in this manuscript:

PAT	Photo-Acoustic Tomography
TR	Time Reversal
ATR	Averaged Time Reversal
DL	Deep Learning
CT	Computed Tomography
CNN	Convolutional Neural Network
PSNR	Peak Signal-to-Noise Ratio
SSIM	Structural Similarity
FDTD	Finite Difference Time Domain
CFL	Courant–Friedrichs–Lewy
MSE	Mean-Squared Error
ASIR	Adaptive Statistical Iterative Reconstruction
HU	Hounsfield unit

Appendix A. Artifacts Correction by Iterative Algorithms

In this appendix, we describe the principles of the averaged time reversal and the adjoint method in more details. Interested readers are referred to [8,14] for the accurate and complete exhibition.

Appendix A.1. Averaged Time Reversal

The averaged time reversal method is proposed in [12] as a remedy to the conventional TR to remove the reflection artifacts in the latter. The essential improvement comes from the introduction of an averaging process to the measured data before reversing the time. The rational, roughly speaking, is that artifacts generated by acoustic reflections have either positive amplitude or negative amplitude, depending on the number of times they touch the boundary. A suitable averaging process along the time direction is thus able to annihilate reflected artifacts with opposite signs. The process can be viewed as a pre-conditioning to the TR. To be a bit more precise, let Λ be the linear operator

$$\Lambda: \qquad p_0 \mapsto u|_{[0,T]\times\partial\Omega} \qquad (A1)$$

where u is the solution of the forward model (1). In other words, Λ sends the initial pressure to the measured data. This is a linear operator that is completed determined by the form of the partial differential equation in (1). If the forward model is discretized, Λ would just be a matrix. Reconstructing the initial pressure from the ultrasound signals amounts to inverting the matrix Λ. Direct inversion is normally impossible due to the large dimensionality of this matrix. Instead, it is shown in [12] that one can introduce an averaged time reversal process A such that the initial pressure p_0 can be reconstructed iteratively from the measured data $u|_{[0,T]\times\partial\Omega}$ by the relation

$$p_0 = \sum_{k=0}^{\infty}(\mathbf{A}\boldsymbol{\Lambda} - \boldsymbol{Id})^k \mathbf{A}(u|_{[0,T]\times\partial\Omega}) \tag{A2}$$

where $\boldsymbol{Id}$ is the identity operator (or identity matrix if the forward model is discretized). This algorithm is known as the averaged time reversal (ATR). It is mathematically proved in [12] that ATR can correct the reflection artifacts caused by conventional time reversal methods. It is one of the baseline methods we used in the paper to compare with the DL reconstruction.

Appendix A.2. Landweber Iteration

The Landweber iteration, also known as the Landweber algorithm, is an algorithm proposed in the 1950s by Landweber [56] to solve ill-posed linear systems of the form $\boldsymbol{\Lambda} x = b$ where $\boldsymbol{\Lambda}$ is a (not necessarily square) matrix. It is a regularization method that can be viewed as iteratively solving the unconstrained optimization problem

$$\min_x \frac{1}{2}\|\boldsymbol{\Lambda} x - b\|_2^2, \tag{A3}$$

which leads to the update scheme

$$x_{k+1} = x_k - \gamma\boldsymbol{\Lambda}^*(\boldsymbol{\Lambda} x_k - b), \qquad k = 0, 1, 2, \ldots \tag{A4}$$

where $\boldsymbol{\Lambda}^*$ is the adjoint operator of $\boldsymbol{\Lambda}$ and γ is a regularization parameter. It is well known the Landweber iteration is convergent if $0 < \gamma < \frac{2}{\sigma_1^2}$ where σ_1 is the largest singular value of $\boldsymbol{\Lambda}$, and it converges to the projection of the true solution on the orthogonal complement of the null space of $\boldsymbol{\Lambda}$, see the analysis in [14] for instance. In our case, the vector x is the discretized version of p_0, and the vector b is the discretized version of the measured data $u|_{[0,T]\times\partial\Omega}$. The iteration exhibits the effect of semi-convergence: it converges before reaching a certain number of iterations but then diverges once beyond. The optimal value of γ and the stopping rule are largely empirical. Some guiding principles exist, but we do not intend to discuss them in this article.

References

1. Xia, J.; Yao, J.; Wang, L.V. Photoacoustic tomography: Principles and advances. *Electromagn. Waves* **2014**, *147*, 1. [CrossRef]
2. Wang, L.V.; Beare, G.K. Breaking the Optical Diffusion Limit: Photoacoustic Tomography. In *Frontiers in Optics*; Optical Society of America: Rochester, NY, USA, 2010; p. FWY2.
3. Yao, J.; Wang, L.V. Photoacoustic tomography: Fundamentals, advances and prospects. *Contrast Media Mol. Imag.* **2011**, *6*, 332–345. [CrossRef] [PubMed]
4. Cox, B.; Beard, P. Photoacoustic tomography with a single detector in a reverberant cavity. *J. Acoust. Soc. Am.* **2009**, *125*, 1426–1436. [CrossRef] [PubMed]
5. Xu, M.; Wang, L.V. Universal back-projection algorithm for photoacoustic computed tomography. *Phys. Rev. E* **2005**, *71*, 016706. [CrossRef] [PubMed]
6. Hristova, Y.; Kuchment, P.; Nguyen, L. Reconstruction and time reversal in thermoacoustic tomography in acoustically homogeneous and inhomogeneous media. *Inverse Probl.* **2008**, *24*, 055006. [CrossRef]
7. Hristova, Y. Time reversal in thermoacoustic tomography—An error estimate. *Inverse Probl.* **2009**, *25*, 055008. [CrossRef]
8. Stefanov, P.; Uhlmann, G. Thermoacoustic tomography with variable sound speed. *Inverse Probl.* **2009**, *25*, 075011. [CrossRef]
9. Stefanov, P.; Uhlmann, G. Thermoacoustic tomography arising in brain imaging. *Inverse Probl.* **2011**, *27*, 045004. [CrossRef]
10. Stefanov, P.; Yang, Y. Thermo and Photoacoustic Tomography with variable speed and planar detectors. *SIAM J. Math. Anal.* **2017**, *49*, 297–310. [CrossRef]

11. Holman, B.; Kunyansky, L. Gradual time reversal in thermo-and photo-acoustic tomography within a resonant cavity. *Inverse Probl.* **2015**, *31*, 035008. [CrossRef]
12. Stefanov, P.; Yang, Y. Multiwave tomography in a closed domain: Averaged sharp time reversal. *Inverse Probl.* **2015**, *31*, 065007. [CrossRef]
13. Nguyen, L.V.; Kunyansky, L.A. A dissipative time reversal technique for photoacoustic tomography in a cavity. *SIAM J. Imag. Sci.* **2016**, *9*, 748–769. [CrossRef]
14. Stefanov, P.; Yang, Y. Multiwave tomography with reflectors: Landweber's iteration. *Inverse Probl. Imag.* **2017**, *11*, 373–401. [CrossRef]
15. Wang, G. A perspective on deep imaging. *IEEE Access* **2016**, *4*, 8914–8924. [CrossRef]
16. Wang, G.; Ye, J.C.; Mueller, K.; Fessler, J.A. Image reconstruction is a new frontier of machine learning. *IEEE Trans. Med. Imag.* **2018**, *37*, 1289–1296. [CrossRef] [PubMed]
17. Zhu, B.; Liu, J.Z.; Cauley, S.F.; Rosen, B.R.; Rosen, M.S. Image reconstruction by domain-transform manifold learning. *Nature* **2018**, *555*, 487. [CrossRef] [PubMed]
18. Antholzer, S.; Haltmeier, M.; Schwab, J. Deep Learning for Photoacoustic Tomography from Sparse Data. *Inverse Probl. Sci. Eng.* **2019**, *27*, 987–1005. [CrossRef] [PubMed]
19. Antholzer, S.; Haltmeier, M.; Nuster, R.; Schwab, J. Photoacoustic image reconstruction via deep learning. In *Photons Plus Ultrasound: Imaging and Sensing 2018*; International Society for Optics and Photonics: Bellingham, WA, USA, 2018; Volume 10494.
20. Antholzer, S.; Schwab, J.; Haltmeier, M. Deep Learning Versus ℓ^1-Minimization for Compressed Sensing Photoacoustic Tomography. In Proceedings of the IEEE International Ultrasonics Symposium (IUS), Kobe, Japan, 22–25 October 2018; pp. 206–212.
21. Hauptmann, A.; Lucka, F.; Betcke, M.M.; Huynh, N.; Adler, J.; Cox, B.T.; Beard, P.C.; Ourselin, S.; Arridge, S.R. Model-Based Learning for Accelerated, Limited-View 3-D Photoacoustic Tomography. *IEEE Trans. Med. Imag.* **2018**, *37*, 1382–1393. [CrossRef] [PubMed]
22. Waibel, D.; Gröhl, J.; Isensee, F.; Kirchner, T.; Maier-Hein, K.; Maier-Hein, L. Reconstruction of initial pressure from limited view photoacoustic images using deep learning. In *Photons Plus Ultrasound: Imaging and Sensing 2018*; International Society for Optics and Photonics: Bellingham, WA, USA, 2018; Volume 10494.
23. Schwab, J.; Antholzer, S.; Nuster, R.; Paltauf, G.; Haltmeier, M. Deep Learning of truncated singular values for limited view photoacoustic tomography. In *Photons Plus Ultrasound: Imaging and Sensing 2019*; International Society for Optics and Photonics: San Francisco, CA, USA, 2019; Volume 10878, p. 1087836.
24. Guan, S.; Khan, A.; Sikdar, S.; Chitnis, P. Fully Dense UNet for 2D sparse photoacoustic tomography artifact removal. *IEEE J. Biomed. Health Inform.* **2019**.[CrossRef]
25. Allman, D.; Reiter, A.; Bell, M.A.L. Photoacoustic source detection and reflection artifact removal enabled by deep learning. *IEEE Trans. Med. Imag.* **2018**, *37*, 1464–1477. [CrossRef]
26. Allman, D.; Reiter, A.; Bell, M. Exploring the effects of transducer models when training convolutional neural networks to eliminate reflection artifacts in experimental photoacoustic images. In *Photons Plus Ultrasound: Imaging and Sensing 2018*; International Society for Optics and Photonics: Bellingham, WA, USA, 2018; Volume 10494, p. 104945H.
27. Kelly, B.; Matthews, T.P.; Anastasio, M.A. Deep learning-guided image reconstruction from incomplete data. *arXiv* **2017**, arXiv:1709.00584.
28. Schwab, J.; Antholzer, S.; Nuster, R.; Haltmeier, M. Real-time photoacoustic projection imaging using deep learning. *arXiv* **2018**, arXiv:1801.06693.
29. Shan, H.; Zhang, Y.; Yang, Q.; Kruger, U.; Kalra, M.K.; Sun, L.; Cong, W.; Wang, G. 3-D convolutional encoder-decoder network for low-dose CT via transfer learning from a 2-D trained network. *IEEE Trans. Med. Imag.* **2018**, *37*, 1522–1534. [CrossRef] [PubMed]
30. You, C.; Yang, Q.; Shan, H.; Gjesteby, L.; Li, G.; Ju, S.; Zhang, Z.; Zhao, Z.; Zhang, Y.; Cong, W.; et al. Structurally-Sensitive Multi-Scale Deep Neural Network for Low-Dose CT Denoising. *IEEE Access* **2018**, *6*, 41839–41855. [CrossRef] [PubMed]
31. Gjesteby, L.; Yang, Q.; Xi, Y.; Shan, H.; Claus, B.; Jin, Y.; De Man, B.; Wang, G. Deep learning methods for CT image-domain metal artifact reduction. In *Developments in X-ray Tomography XI*; International Society for Optics and Photonics: San Diego, CA, USA, 2017; Volume 10391, p. 103910W.

32. Gjesteby, L.; Shan, H.; Yang, Q.; Xi, Y.; Claus, B.; Jin, Y.; De Man, B.; Wang, G. Deep Neural Network for CT Metal Artifact Reduction with a Perceptual Loss Function. In Proceedings of the 5th International Conference on Image Formation in X-Ray Computed Tomography, Salt Lake City, UT, USA, 20–23 May 2018.

33. You, C.; Li, G.; Zhang, Y.; Zhang, X.; Shan, H.; Ju, S.; Zhao, Z.; Zhang, Z.; Cong, W.; Vannier, M.W.; et al. CT Super-resolution GAN Constrained by the Identical, Residual, and Cycle Learning Ensemble (GAN-CIRCLE). *IEEE Trans. Med. Imag.* **2019**. [CrossRef]

34. Lyu, Q.; You, C.; Shan, H.; Zhang, Y.; Wang, G. Super-resolution MRI and CT through GAN-circle. In *Developments in X-Ray Tomography XI*; International Society for Optics and Photonics: San Diego, CA, USA, 2019.

35. Shan, H.; Padole, A.; Homayounieh, F.; Kruger, U.; Khera, R.D.; Nitiwarangkul, C.; Kalra, M.K.; Wang, G. Competitive performance of a modularized deep neural network compared to commercial algorithms for low-dose CT image reconstruction. *Nat. Mach. Intell.* **2019**, *1*, 269–276. [CrossRef]

36. Kuchment, P.; Kunyansky, L. Mathematics of thermoacoustic tomography. *Eur. J. Appl. Math.* **2008**, *19*, 191–224. [CrossRef]

37. Ammari, H.; Bossy, E.; Jugnon, V.; Kang, H. Mathematical modeling in photoacoustic imaging of small absorbers. *SIAM Rev.* **2010**, *52*, 677–695. [CrossRef]

38. Ammari, H.; Asch, M.; Bustos, L.G.; Jugnon, V.; Kang, H. Transient wave imaging with limited-view data. *SIAM J. Imag. Sci.* **2011**, *4*, 1097–1121. [CrossRef]

39. Acosta, S.; Montalto, C. Multiwave imaging in an enclosure with variable wave speed. *Inverse Probl.* **2015**, *31*, 065009. [CrossRef]

40. Ronneberger, O.; Fischer, P.; Brox, T. U-net: Convolutional networks for biomedical image segmentation. In *International Conference on Medical Image Computing and Computer-Assisted Intervention*; Springer: Cham, Munich, Germany, 2015; pp. 234–241.

41. He, K.; Zhang, X.; Ren, S.; Sun, J. Deep residual learning for image recognition. In Proceedings of the IEEE Conference on Computer Vision and Pattern Recognition, Las Vegas, NV, USA, 26 June–1 July 2016; pp. 770–778.

42. Glorot, X.; Bordes, A.; Bengio, Y. Deep sparse rectifier neural networks. In Proceedings of the Fourteenth International Conference on Artificial Intelligence and Statistics, Fort Lauderdale, FL, USA, 11–13 April 2011; pp. 315–323.

43. Wang, Z.; Bovik, A.C.; Sheikh, H.R.; Simoncelli, E.P. Image quality assessment: From error visibility to structural similarity. *IEEE Trans. Image Process.* **2004**, *13*, 600–612. [CrossRef] [PubMed]

44. Zhao, H.; Gallo, O.; Frosio, I.; Kautz, J. Loss functions for image restoration with neural networks. *IEEE Trans. Comput. Imag.* **2016**, *3*, 47–57. [CrossRef]

45. Kingma, D.P.; Ba, J. Adam: A method for stochastic optimization. *arXiv* **2014**, arXiv:1412.6980.

46. Rumelhart, D.E.; Hinton, G.E.; Williams, R.J. Learning representations by back-propagating errors. *Cognit. Model.* **1988**, *5*, 1. [CrossRef]

47. Yang, Q.; Kalra, M.K.; Padole, A.; Li, J.; Hilliard, E.; Lai, R.; Wang, G. Big data from CT scanning. *JSM Biomed. Imag.* **2015**, *2*, 1003.

48. McCollough, C.H.; Bruesewitz, M.R.; Kofler Jr, J.M. CT dose reduction and dose management tools: Overview of available options. *Radiographics* **2006**, *26*, 503–512. [CrossRef]

49. Qian, J.; Stefanov, P.; Uhlmann, G.; Zhao, H. An efficient Neumann series-based algorithm for thermoacoustic and photoacoustic tomography with variable sound speed. *SIAM J. Imag. Sci.* **2011**, *4*, 850–883. [CrossRef]

50. Treeby, B.E.; Zhang, E.Z.; Cox, B.T. Photoacoustic tomography in absorbing acoustic media using time reversal. *Inverse Probl.* **2010**, *26*, 115003. [CrossRef]

51. Belhachmi, Z.; Glatz, T.; Scherzer, O. A direct method for photoacoustic tomography with inhomogeneous sound speed. *Inverse Probl.* **2016**, *32*, 045005. [CrossRef]

52. Javaherian, A.; Holman, S. Direct quantitative photoacoustic tomography for realistic acoustic media. *Inverse Probl.* **2019**. [CrossRef]

53. Javaherian, A.; Holman, S. A continuous adjoint for photo-acoustic tomography of the brain. *Inverse Probl.* **2018**, *34*, 085003. [CrossRef]

54. Treeby, B.E.; Cox, B.T. k-Wave: MATLAB toolbox for the simulation and reconstruction of photoacoustic wave-fields. *J. Biomed. Opt.* **2010**, *15*, 021314. [CrossRef] [PubMed]

55.   Paszke, A.; Gross, S.; Chintala, S.; Chanan, G.; Yang, E.; DeVito, Z.; Lin, Z.; Desmaison, A.; Antiga, L.; Lerer, A.   Automatic differentiation in PyTorch.   In Proceedings of the NIPS 2017 Autodiff Workshop, Long Beach, CA, USA, 9 December 2017.
56.   Landweber, L. An iteration formula for Fredholm integral equations of the first kind. *Am. J. Math.* **1951**, *73*, 615–624. [CrossRef]

Article

Reconstruction of Photoacoustic Tomography Inside a Scattering Layer Using a Matrix Filtering Method

Wei Rui [1], Zhipeng Liu [1], Chao Tao [2,*] and Xiaojun Liu [1,*]

[1] Key Laboratory of Modern Acoustics, Department of Physics and Collaborative Innovation Center of Advanced Microstructures, Nanjing University, Nanjing 210093, China; dz1522101@smail.nju.edu.cn (W.R.); mg1722085@smail.nju.edu.cn (Z.L.)

[2] Shenzhen Research Institute of Nanjing University, Shenzhen 51800, China

* Correspondence: taochao@nju.edu.cn (C.T.); liuxiaojun@nju.edu.cn (X.L.)

Received: 29 April 2019; Accepted: 17 May 2019; Published: 20 May 2019

Abstract: Photoacoustic (PA) tomography (PAT) has potential for use in brain imaging due to its rich optical contrast, high acoustic resolution in deep tissue, and good biosafety. However, the skull often poses challenges for transcranial brain imaging. The skull can cause severe distortion and attenuation of the phase and amplitude of PA waves, which leads to poor resolution, low contrast, and strong noise in the images. In this study, we propose an image reconstruction method to recover the PA image insider a skull-like scattering layer. This method reduces the scattering artifacts by combining a correlation matrix filter and a time reversal operator. Both numerical simulations and PA imaging experiments demonstrate that the proposed approach effectively improves the image quality with less speckle noise and better signal-to-noise ratio. The proposed method may improve the quality of PAT in a complex acoustic scattering environment, such as transcranial brain imaging.

Keywords: photoacoustic imaging; correlation matrix filter; time reversal operator

1. Introduction

Photoacoustic (PA) tomography (PAT) is a noninvasive and nonionizing imaging technique that combines rich optical contrast and high ultrasonic resolution in deep tissue [1–4]. During the PAT imaging process, the biological tissue is irradiated by a laser pulse, and then optical absorbers in the tissue absorb electromagnetic energy and generate ultrasound due to rapid thermoelastic expansion. The ultrasound signal is then detected by the ultrasound transducer array arranged outside the tissue. Finally, images with optical absorption contrasts are extracted from the detected PA signal using a reconstruction algorithm. PAT breaks through the optical diffraction limit and obtains images of tissue at depths of 5–6 cm with good acoustic spatial resolution. Therefore, PAT has application potential for high-quality in vivo vascular structure imaging [5], investigations of breast tumor tissue [6–9], vasculature visualization [10,11], small animal whole-body imaging [12,13], osteoarthritis assessment [14,15], drug delivery monitoring [16], and hemodynamic functional imaging [17].

One promising application of PAT is high-resolution in vivo brain imaging, which is important in neurophysiology, neuropathology, and neurotherapy. In comparison with the often-used brain imaging modalities, including X-ray computerized tomography (CT) and magnetic resonance imaging (MRI), PAT is inexpensive and has good spatial resolution and temporal resolution [1,2]. X-ray CT involves exposure to ionizing radiation, but PAT is nonionizing and safer. PAT can provide structural images as well as functional images of the brain.

The classical image reconstruction algorithms used in PAT are usually based on the principle of coherent beamforming, which relies on an assumption of homogeneous media. In a scene with strong scattering, the coherence in signals is destroyed by the randomness of scattering, leading to a markedly decreased penetration depth and significant degradation in image quality [18]. This limitation hinders the application of PAT in a wider range of scenarios, such as transcranial PA brain imaging within the skull bone. The skull bone has much higher acoustic impedance than soft tissue. A severe acoustic impedance mismatch could cause amplitude and phase distortion, and decrease the resolution and contrast of the transcranial imaging, significantly reducing the brain image quality [19–22]. Traditional approaches to overcome the limitation of acoustic scattering have mostly relied on a priori knowledge of the properties of tissue inhomogeneity or need to involve additional acoustic measurements, evidenced in the statistical reconstruction method [23], coherence factor optimization [24], and the interferometry method [25]. For a more general situation, a universal reconstruction algorithm with better performance is still required for acquiring high-quality images in scattering media.

In this study, we created an image reconstruction method that combines a correlation matrix filter and a time reversal operator using an annular transducer array. After dividing the circular array into multiple sections, each sector array can be approximated as a linear array via phase and amplitude compensation. Then, using the correlation characteristics of the direct waves, a correlation full-matrix filter can be applied to separate the direct waves from the scattering waves [26–29]. Finally, the imaging results of each section based on the time reversal method are superimposed to further improve the image quality. Both numerical simulation and phantom experiments were employed to validate the proposed method and examine its potential for extracting PA images inside a skull-like scattering layer.

2. Methods

As shown in Figure 1, an imaging target is placed in the region of interest (ROI), which is surrounded by a skull-like scattering layer. Under the illumination of the pulse laser, the optical absorber will radiate ultrasound waves to the surrounding medium as a result of the PA effect. After the acoustic waves propagate through the random scattering layer, a passive ultrasound annular array with 576 elements detects the ultrasound signal. The annular array is then divided into N_S sectors for data processing and imaging. Each sector has $N = 45$ elements and can be approximated as a linear array so that the filtered imaging method proposed in our previous studies can be applied for the image reconstruction [26,27]. For each sector array, the recorded N-channel signals are arranged in a vector $\mathbf{H}(t) = [H_1(t), H_2(t), \ldots, H_n(t), \ldots, H_N(t)]$ where $n = 1, 2, \ldots, N$.

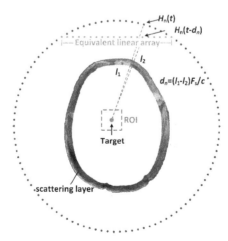

Figure 1. The schematic of the scenario considered in this study. For any target placed in the region of interest (ROI) within a scattering layer, its propagation distances to the nth transducers of a sector array, and the equivalent linear array are l_1 and l_2, respectively. The corresponding delay is then given by $d_n = (l_1 - l_2)F_s/c$, where c is the sound speed and F_s is the sampling frequency.

The sector array can be approximated to a linear array by calculating the corresponding delay. For any imaging target, its propagation distances to the nth transducers of the two arrays are l_1 and l_2, respectively. The corresponding delay is $d_n = (l_1 - l_2)F_s/c$, where c is the sound speed and F_s is the sampling frequency. The time domain signals for the linear array are then $\mathbf{J}(t) = [H_1(t - d_1), H_2(t - d_2), \dots, H_n(t - d_n), \dots, H_N(t - d_N)]$. The signals can be written in their frequency domain by applying a Fourier transform to $\mathbf{P}(T, f) = [P_1(T, f), P_2(T, f), \dots, P_n(T, f), \dots, P_N(T, f)]$, where $P_n(T, f)$ is the short-time Fourier transform of the nth channel signal in the time window $[T - \Delta t/2, T + \Delta t/2]$, T is the time of flight, and Δt is the width of the window. $\mathbf{P}(T, f)$ contains both the direct wave $\mathbf{P}^D(T, f)$ and the scattering wave $\mathbf{P}^S(T, f)$. When the concentration of the scatterers is sufficiently high, the scattering wave occupies a dominant position and the imaging quality is greatly degraded due to the randomness of the scattering process.

To reduce the influence of the randomly scattering, a matrix filtering method can be applied to extract the direct wave from the received signals of a linear array [26]. Firstly, a response matrix can be constructed as:

$$\mathbf{K} = \mathbf{PP}^T = \underbrace{\mathbf{P}^D(\mathbf{P}^D)^T}_{\text{Coherence,}\mathbf{K}^C} + \underbrace{[\mathbf{P}^D(\mathbf{P}^S)^T + \mathbf{P}^S(\mathbf{P}^D)^T + \mathbf{P}^S(\mathbf{P}^S)^T]}_{\text{Random,}\mathbf{K}^R} \tag{1}$$

Depending on whether or not contain the term $\mathbf{P}^S$, the matrix $\mathbf{K}$ can be divided into two parts: $\mathbf{K}^R$ and $\mathbf{K}^C$, where $\mathbf{K}^R$ is a random term due to the random nature of the scatterer distribution and the scattering paths and $\mathbf{K}^C$ is independent of the scattering paths and shows a deterministic relation of the phase between its elements along the antidiagonal:

$$\beta_q = \frac{k^C_{m-q,m+q}}{k^C_{m,m}} = \exp[jk(qw)^2/Z] \tag{2}$$

where k is the wave number, w is the pitch size, and Z is the depth of the time window. The phase relation along the antidiagonal only depends on the channel position qw, indicating that the direct waves from the optical absorbers manifest themselves as a particular coherence on the antidiagonals of the matrix.

Using the coherent property, we can extract the direct wave by the following steps. Firstly, the original matrix $\mathbf{K}$ is rotated 45° counterclockwise so that the coherent property occurs on the columns. The elements are divided into two parts based on their symmetry characteristic and are expanded to two new matrices: $\mathbf{A}_1$ and $\mathbf{A}_2$ [26]. The matrices $\mathbf{A}_1$ and $\mathbf{A}_2$ are called indifferently $\mathbf{A}$ as they are filtered in the same way. A filtering matrix $\mathbf{F} = \mathbf{CC}^\dagger$ is then applied to extract the coherent part from matrix $\mathbf{A}$ as: $\mathbf{A}^F = \mathbf{FA} = \mathbf{FA}^C + \mathbf{FA}^R$, where $\mathbf{C} = [\exp(jky_1^2/2Z), \dots, \exp(jky_u^2/2Z), \dots, \exp(jky_{N_A}^2/2Z)]$ with $u = 1, 2, \dots, N_A$, and $y_u = [x_m + x_n + (N_A - 1)w]$ is the characteristic spacing of direct waves. The coherent part remains unchanged during the filtering process, as $\mathbf{FA}^C = \mathbf{A}^C$ and the random part decreases by a factor $\sqrt{(N + 1)/2}$. Finally, a filtered matrix $\mathbf{K}^F$ with the original coordinate can be obtained by applying the inverse process of the first step to matrices $\mathbf{A}_1^F$ and $\mathbf{A}_2^F$.

Using a time reversal operator, we can recover the image of the optical absorbers from the matrix $\mathbf{K}^F$ corresponding to the single sector area [28,29]. The singular value decomposition is applied to the matrix $\mathbf{K}^F = \mathbf{U}^F\mathbf{\Lambda}^F\mathbf{V}^{F\dagger}$, and each optical absorber is mainly associated with one non-zero singular λ_i^F contained in diagonal matrix $\mathbf{\Lambda}$ [30]. Therefore, by numerically back-propagating the singular vector, the image of the ith singular valued at a depth of $Z = cT$ can be achieved by the time reversal operator $\mathbf{I}_i = \lambda_i^F|\mathbf{G} \times \mathbf{V}|$, where $\mathbf{G}$ with the component $g_m = \exp(jkr_m)/\sqrt{r_m}$ is the propagating operator between the array and the focal plane in a homogeneous media, and r_m is the distance between the mth element in the array and the point in the focal plane. By superimposing the imaging results of multiple sectors, we can further improve the imaging quality and reduce the influence of the inhomogeneous aberrating effect of the scattering layer. For the wanted imaging point A, we calculated the image value I_m^A for all the linear arrays with $m = 1, 2, \dots, N_s$. Taking the first sector area as an example, the image value for points corresponding to the array can be obtained by applying the time reversal operator. If point A is within the imaging area, we can find two imaging points, B and C, closest to point A in the focal plane and then perform a weighted evaluation based on the distance from point A to points B and C:

$$I_1^A = \frac{I_1^B d_{AC} + I_1^C d_{AB}}{d_{AC} + d_{AB}} \tag{3}$$

where d_{AC} and d_{AB} represent the distances between A, B and A, C, respectively. If point A is outside the imaging area, the imaging value is then recorded as 0. When we calculate the imaging value of point A for all sector areas, its final value is the result of taking the summation and average to all the non-zero values: $I^A = \sum_{m=1}^{N_s} I_m^A / (N_s - N_z)$, where N_Z is the number of zero values. As such, we can fully use the valid information received by all the transducers of the annular array for image reconstruction of each point in the ROI.

3. Results

3.1. Numerical Simulation

Numerical simulations were used to validate the proposed method. The simulation is depicted in Figure 2. The radius of the annular array was set to $R_a = 40$ mm, and there were $N = 576$ elements in the array. The acoustic scattering layer consisted of 120 acoustic scatterers with a diameter of $d = 0.8$ mm. The inner and outside radius of the scattering layer were $R_1 = 10$ mm and $R_2 = 15$ mm, respectively. The sound velocity and the density of these scatterers were 5200 m/s and 7,870 kg/m^3, respectively. The sound velocity of the surrounding medium was 1500 m/s, the density of which was 1,000 kg/m^3, which is close to that of soft tissues or water. Therefore, the scatterers had a severe impedance mismatch with the background media, which resulted in strong scattering. The imaging targets were five optical absorbers with a diameter of $d = 0.8$ mm in the ROI within the scattering layer. The coordinates of these targets were (−2,−2), (−2,2), (2,−2), (2,2), and (0,0). Under the illumination of the laser pulse, these optical absorbers in the scattering layer emitted ultrasonic pulses with a central frequency of 2.0 MHz and a −3 dB bandwidth of 1.26–2.68 MHz to the surrounding medium. The PA signals were subsequently detected by the annular array enclosing the region of interest. The PA signals awee sampled with a frequency of 40 Mhz. The imaging targets had the same sound velocity and the same density as the scatterers.

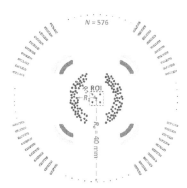

Figure 2. The schematic of the scenario of the simulation. Five identical optical absorbers with a diameter of $d = 0.8$ mm are placed in the ROI within a scattering layer. The inner and outside radius of the scattering layer are $R_1 = 10$ mm and $R_2 = 15$ mm, respectively. The ultrasound signal is detected by an annular array with a radius of $R_a = 40$ mm.

Figure 3 depicts the results of the simulation with the scattering layer. Figure 3a illustrates the typical PA signals detected by one transducer. The measured broadband signal was partly overwhelmed by the noise and the amplitude was unstable. The PA signals detected by all transducers are shown in Figure 3b to visualize the effect of the scattering layer on the received signals, where the gray dotted line corresponds to the results in Figure 3a. The coherence of the signals was broken by the randomness of the scattering. The presence of a scattering layer inevitably causes a drop in the image quality.For the sake of comparison, Figure 3c illustrates the imaging results reconstructed by the classic delay-and-sum (DAS) beamforming method. As shown, for the DAS beamforming method, the image reconstructed in the case with a scattering layer (Figure 3c) suffered from low resolution, and the background showed strong noise. As a result, the contrast between the targets and the background decreased distinctly. Figure 3d depicts the image recovered by the time reversal method combined with a correlation matrix filter. The image shows good contrast, as the background noise has been significantly reduced after filtering.

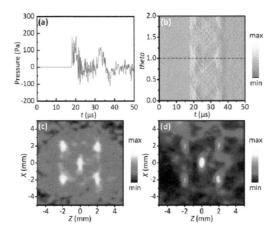

Figure 3. Simulative results. (**a**) The detected photoacoustic (PA) signals by a single transducer. (**b**) The detected PA signals by all transducers. (**c**) The image reconstructed by the delay and sum (DAS) beamforming method. (**d**) The image reconstructed by the proposed matrix filtering method.

We used the signal-to-noise ratio (SNR) and the full width at half maximum (FWHM) of the targets' image to quantify the image quality. The SNR is defined as the ratio of the mean signal intensity in the signal area and the mean noise intensity in the background area. We defined the area inside the circles with a diameter of 0.8 mm (equal to the actual size of the targets) at the locations of the five targets as the signal area and the area outside as the background area. The averaged FWHM along the X and Z axes of all targets was calculated to quantify the degree of the distortion. For the results in Figure 3c, the SNR of the image was 17.2 dB, and the averaged FWHM along the X and Z axes were 1.42 mm and 2.6 mm, respectively. The difference between the FWHM along the X and Z axes were caused by the non-uniform distribution of the scatterers in the scattering layer. The SNR of the image in Figure 3d increased to 21.0 dB due to the matrix filter. In addition, the FWHM along the X and Z axes were 0.87 mm and 1.63 mm, respectively, which are both closer to the actual size of the targets. The results demonstrate that the matrix filtering method can more accurate depict the shape and position of the targets in the ROI.

3.2. Experimental Study

We also performed an experiment on a vessel shape source to validate the performance of the proposed method. A schematic diagram of the experimental system is provided in Figure 4a. A vessel-shaped source placed in the ROI was irradiated by a Q-switched Nd: YAG pulse laser with a wavelength of 532 nm. The pulse width was about 8 ns. The ultrasonic waves were detected by a 5 MHz central frequency line focused ultrasonic transducer (V310-SM, PANAMETRICS-NDT) with a −6 dB bandwidth of 4.4 MHz, then amplified (SA-230F5, NF) and sampled (PCI-5105, NI) at a sampling rate of 60 Mhz. Under the control of a rotary stepper motor, the transducer scanned clockwise around the sample with a radius of 4.3 cm and a step of 2°. It is equivalent to an annular array with 180 elements.

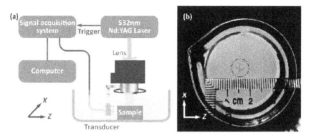

Figure 4. PA imaging of a vessel-shaped source experiment. (**a**) Schematic diagram of the PA imaging experimental setup. (**b**) Photo of the sample.

Figure 4b is a picture of the imaging sample. A vessel-shaped source made of hair was embedded in agar for fixing, with an agar to water ratio of about 1.2%, and the sound velocity was measured as 1,460 m/s at 25 °C. To mimic the effect of the skull on brain imaging, we placed the sample in a glass beaker. The glass has a thickness of about 1.0 mm, and the corresponding speed of sound and density were 2,730 m/s and 2,500 kg/m^3, respectively. The acoustic impedance of the glass beaker was much higher than that of the agar and the surrounding water. It caused strong acoustic scattering of the PA waves, as would the skull. Serious acoustic impedance mismatch between the glass beaker and the surrounding media could result in a significant drop in image quality.

For comparison, the images recovered by the DAS method and the proposed method in the cases with and without the beaker are shown in Figure 5. Figure 5a,b show the imaging results of the DAS method and the proposed method without the beaker. In both cases, the pattern is clear enough that the shape of the source is distinguishable.

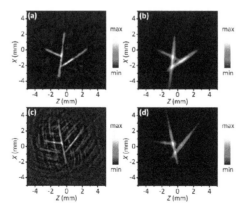

Figure 5. Reconstructed images of the vessel shape source in the ROI. Reconstructed image of the sample without the beaker (**a**) by the DAS beamforming method and (**b**) by the proposed matrix filtering method. Reconstructed image of the sample with the beaker (**c**) by the DAS beamforming method and (**d**) by the proposed matrix filtering method.

However, for the sample in a beaker, the image recovered by the DAS method (Figure 5c) shows low contrast and strong background noise. The shape of the source is hard to distinguish. In the imaging result produced by our proposed method, we obtained a clear pattern of the vessel shape source with high contrast, which is similar to the result in Figure 5b. We conclude that the proposed method can effectively reduce the degradation of imaging quality caused by the beaker, and the experiments validated the performance of the proposed method.

4. Discussion and Conclusions

In summary, we developed an image reconstruction method for PAT. This method combines a correlation matrix filter and a time reversal operator. We applied this method to reconstruct PA images inside a scattering layer. Both numerical simulation and phantom experiments were used to examine the performance of the proposed method. The simulation quantitively validated that the proposed imaging method provides higher imaging quality in complex scattering media compared with the classic beamforming method. The advantages are reflected in better a SNR and FWHM.

Though demonstrating potential in preclinical and clinical applications, PAT is still in its early age of development. Among the several limitations preventing PAT's application, the main one is the amplitude and phase distortion of the received signals caused by the randomly distributed scatterers. In this research, we proposed an image reconstruction method to reduce the negative effect of the scattering waves by separating the direct waves from the received signals. With the filtered signals, PA images can be optimized, especially the resolution and the contrast between the targets and the background. This work might be valuable for understanding the physics of the interaction of sound in the complex media. The proposed method may potentially be used for acquiring high-quality images of the brain through the skull.

In this study, the experiments were performed by a point-by-point scanning with a single ultrasound transducer. Point-by-point scanning is not fast enough to meet the imaging requirements in pre-clinical or clinical scenarios. This problem can be solved by using a circular ultrasound array, which is what we are preparing to study in our next work. Additionally, as future work, we will evaluate the proposed method in an in vivo study on a small animal.

Author Contributions: R.W. designed the simulations, employed the experiments, analyzed the results, and wrote the paper. Z.L. assisted in both the experiments employment and writing of the paper.

Funding: This research was funded by the National Key R&D Program of China (2016YFC0102300), National Natural Science Foundation of China (NSFC) (11834008, 11874217), Fundamental Research Funds for the Central Universities and Natural Science Foundation of Jiangsu Province (No. BK 20181077).

Conflicts of Interest: The authors declare no conflict of interest.

References

1. Wang, L.V. Multiscale photoacoustic microscopy and computed tomography. *Nat. Photonics* **2009**, *3*, 503. [CrossRef] [PubMed]
2. Zhang, C.; Zhou, Y.; Li, C.; Wang, L.V. Slow-sound photoacoustic microscopy. *Appl. Phys. Lett.* **2013**, *102*, 163702. [CrossRef] [PubMed]
3. Wang, L.V.; Hu, S. Photoacoustic Tomography: In Vivo Imaging from Organelles to Organs. *Science* **2012**, *335*, 1458–1462. [CrossRef] [PubMed]
4. Wang, L.V.; Yao, J.J. A practical guide to photoacoustic tomography in the life sciences. *Nat. Methods* **2016**, *13*, 627–638. [CrossRef]
5. Wang, X.D.; Pang, Y.J.; Ku, G.; Xie, X.Y.; Stoica, G.; Wang, L.V. Noninvasive laser-induced photoacoustic tomography for structural and functional in vivo imaging of the brain. *Nat. Biotechnol* **2003**, *21*, 803–806. [CrossRef] [PubMed]
6. Kang, J.; Kim, E.-K.; Kwak, J.Y.; Yoo, Y.; Song, T.-K.; Chang, J.H. Optimal laser wavelength for photoacoustic imaging of breast microcalcifications. *Appl. Phys. Lett.* **2011**, *99*, 153702. [CrossRef]
7. Olafsson, R.; Bauer, D.R.; Montilla, L.G.; Witte, R.S. Real-time, contrast enhanced photoacoustic imaging of cancer in a mouse window chamber. *Opt. Express* **2010**, *18*, 18625–18632. [CrossRef]

8.	Staley, J.; Grogan, P.; Samadi, A.K.; Cui, H.Z.; Cohen, M.S.; Yang, X.M. Growth of melanoma brain tumors monitored by photoacoustic microscopy. *J. Biomed. Opt.* **2010**, *15*, 3. [CrossRef]
9.	Galanzha, E.I.; Shashkov, E.V.; Kelly, T.; Kim, J.W.; Yang, L.L.; Zharov, V.P. In vivo magnetic enrichment and multiplex photoacoustic detection of circulating tumour cells. *Nat. Nanotechnol.* **2009**, *4*, 855–860. [CrossRef] [PubMed]
10.	Dima, A.; Ntziachristos, V. Non-invasive carotid imaging using optoacoustic tomography. *Opt. Express* **2012**, *20*, 25044–25057. [CrossRef]
11.	Hu, S.; Wang, L.V. Photoacoustic imaging and characterization of the microvasculature. *J. Biomed. Opt.* **2010**, *15*, 011101. [CrossRef]
12.	Brecht, H.P.; Su, R.; Fronheiser, M.; Ermilov, S.A.; Conjusteau, A.; Oraevsky, A.A. Whole-body three-dimensional optoacoustic tomography system for small animals. *J. Biomed. Opt.* **2009**, *14*, 8. [CrossRef] [PubMed]
13.	Xia, J.; Wang, L.V. Small-Animal Whole-Body Photoacoustic Tomography: A Review. *IEEE Trans. Biomed. Eng.* **2014**, *61*, 1380–1389.
14.	Sun, Y.; Sobel, E.S.; Jiang, H.B. First assessment of three-dimensional quantitative photoacoustic tomography for in vivo detection of osteoarthritis in the finger joints. *Med. Phys.* **2011**, *38*, 4009–4017. [CrossRef]
15.	Xiao, J.Y.; Yao, L.; Sun, Y.; Sobel, E.S.; He, J.S.; Jiang, H.B. Quantitative two-dimensional photoacoustic tomography of osteoarthritis in the finger joints. *Opt. Express* **2010**, *18*, 14359–14365. [CrossRef]
16.	Rajian, J.R.; Fabiilli, M.L.; Fowlkes, J.B.; Carson, P.L.; Wang, X. Drug delivery monitoring by photoacoustic tomography with an ICG encapsulated double emulsion. *Opt. Express* **2011**, *19*, 14335–14347. [CrossRef]
17.	Yang, S.; Xing, D.; Zhou, Q.; Xiang, L.; Lao, Y. Functional imaging of cerebrovascular activities in small animals using high-resolution photoacoustic tomography. *Med. Phys.* **2007**, *34*, 3294–3301. [CrossRef]
18.	Robert, J.L.; Fink, M. Green's function estimation in speckle using the decomposition of the time reversal operator: Application to aberration correction in medical imaging. *J. Acoust. Soc. Am.* **2008**, *123*, 866–877. [CrossRef] [PubMed]
19.	Kneipp, M.; Turner, J.; Estrada, H.; Rebling, J.; Shoham, S.; Razansky, D. Effects of the murine skull in optoacoustic brain microscopy. *J. Biophotonics* **2016**, *9*, 117–123. [CrossRef]
20.	Chen, W.T.; Tao, C.; Liu, X.J. Artifact-free imaging through a bone-like layer by using an ultrasonic-guided photoacoustic microscopy. *Opt. Lett.* **2019**, *44*, 1273–1276. [CrossRef]
21.	Jin, X.; Li, C.H.; Wang, L.V. Effects of acoustic heterogeneities on transcranial brain imaging with microwave-induced thermoacoustic tomography. *Med. Phys.* **2008**, *35*, 3205–3214. [CrossRef]
22.	Xu, M.H.; Wang, L.V. Photoacoustic imaging in biomedicine. *Rev. Sci. Instrum.* **2006**, *77*, 22.
23.	Dean-Ben, X.L.; Ntziachristos, V.; Razansky, D. Statistical optoacoustic image reconstruction using a-priori knowledge on the location of acoustic distortions. *Appl. Phys. Lett.* **2011**, *98*, 171110. [CrossRef]
24.	Yoon, C.; Kang, J.; Yoo, Y.; Song, T.K.; Chang, J.H. Enhancement of photoacoustic image quality by sound speed correction: Ex vivo evaluation. *Opt. Express* **2012**, *20*, 3082–3090. [CrossRef]
25.	Yin, J.; Tao, C.; Liu, X. Photoacoustic tomography based on the Green's function retrieval with ultrasound interferometry for sample partially behind an acoustically scattering layer. *Appl. Phys. Lett.* **2015**, *106*, 234101. [CrossRef]
26.	Rui, W.; Tao, C.; Liu, X.J. Imaging acoustic sources through scattering media by using a correlation full-matrix filter. *Sci. Rep.* **2018**, *8*, 15611. [CrossRef]
27.	Rui, W.; Tao, C.; Liu, X.J. Photoacoustic imaging in scattering media by combining a correlation matrix filter with a time reversal operator. *Opt. Express* **2017**, *25*, 22840–22850. [CrossRef]
28.	Aubry, A.; Derode, A. Random Matrix Theory Applied to Acoustic Backscattering and Imaging In Complex Media. *Phys. Rev. Lett.* **2009**, *102*, 4. [CrossRef]
29.	Aubry, A.; Derode, A. Detection and imaging in a random medium: A matrix method to overcome multiple scattering and aberration. *J. Appl. Phys.* **2009**, *106*, 19. [CrossRef]
30.	Aubry, A.; Derode, A. Singular value distribution of the propagation matrix in random scattering media. *Wave Random Complex* **2010**, *20*, 333–363. [CrossRef]

Article

3D Photoacoustic Tomography System Based on Full-View Illumination and Ultrasound Detection

Mingjian Sun [1,2,*], **Depeng Hu** [1], **Wenxue Zhou** [1], **Yang Liu** [2], **Yawei Qu** [2] **and Liyong Ma** [1]

[1] Department of Control Science and Engineering, Harbin Institute of Technology, Weihai 264209, China; wshdphit@163.com (D.H.); zwxhiter@163.com (W.Z.); maly@hitwh.edu.cn (L.M.)

[2] Department of Control Science and Engineering, Harbin Institute of Technology, Harbin 150000, China; liuyang_6Y6@163.com (Y.L.); qyw198553@163.com (Y.Q.)

* Correspondence: sunmingjian@hit.edu.cn; Tel.: +86-183-6318-0812

Received: 25 March 2019; Accepted: 6 May 2019; Published: 9 May 2019

Abstract: A 3D photoacoustic computed tomography (3D-PACT) system based on full-view illumination and ultrasound detection was developed and applied to 3D photoacoustic imaging of several phantoms. The system utilized an optics cage design to achieve full-view uniform laser illumination and completed 3D scanning with the rotation of a dual-element transducer (5 MHz) and the vertical motion of imaging target, which obtains the best solution in the mutual restriction relation between cost and performance. The 3D-PACT system exhibits a spatial resolution on the order of 300 μm, and the imaging area can be up to 52 mm in diameter. The transducers used in the system provides tomography imaging with large fields of view. In addition, the coplanar uniform illumination and acoustic detection configuration based on a quartz bowl greatly enhances the efficiency of laser illumination and signal detection, making it available for use on samples with irregular surfaces. Performance testing and 3D photoacoustic experiments on various phantoms verify that the system can perform 3D photoacoustic imaging on targets with complex surfaces or large sizes. In future, efforts will be made to achieve full-body 3D tomography of small animals and a multimodal 3D imaging system.

Keywords: 3D photoacoustic tomography; full-view illumination and ultrasound detection; photoacoustic coplanar; quartz bowl

---

1. Introduction

Photoacoustic imaging (PAI) is a non-invasive and non-ionized multimodal biomedical imaging method based on transient thermoelastic effects of the biological tissue. The principle is that the energy of pulsed laser light deposited in biological tissue during the process of laser absorption is converted into an ultrasonic signal which is called the photoacoustic signal through instantaneous thermoelastic expansion, and the ultrasonic transducer can receive the signal that carries information about the properties of laser absorption in biological tissue [1–7]. Through the corresponding signal processing and image reconstruction algorithms, a photoacoustic image reflecting the internal structure and function of the tissue can be obtained. Photoacoustic imaging effectively overcomes the limitations of existing pure optical imaging and pure ultrasound imaging, the contrast is based on the absorption of laser light during the photoacoustic excitation period, while the resolution is derived from the ultrasonic detection during the photoacoustic emission period. Photoacoustic imaging breaks through the limit depth of diffusion of high-resolution optical imaging (about 1 mm), and uses laser-generated ultrasonic waves as a carrier to obtain optical absorption information of tissue [8,9]. It is mainly suitable for the tissues whose acoustic properties are uniform but optical properties are not. Therefore, photoacoustic imaging combines the advantages of high contrast characteristics of optical imaging and high penetration characteristics of ultrasonic imaging, and can provide photoacoustic images with high optical contrast and high acoustic resolution at a deeper imaging depth [10]. As a new generation of biomedical imaging technology, photoacoustic imaging can effectively realize the structure and functional imaging of biological tissue, which provides an important method for studying the morphological structure, physiological characteristics, pathological characteristics, and metabolic functions of biological tissues [11–13].

Scientific research on 3D photoacoustic imaging of small animals has become increasingly popular in recent years. There have been many 3D photoacoustic tomography (PACT) systems using different photoacoustic coupling methods, optical transmission models [14], and ultrasonic detection designs [15–20]. However, due to the limited detection angle of ultrasonic transducers, linear array PACT [16–19], spherical array PACT [20,21], arc-shaped array PACT [22], and the half-ring multispectral optoacoustic tomography (MSOT) [23] generally need to obtain complete photoacoustic signal via spatial multiplexing. In photoacoustic tomography, uniform illumination helps to obtain more comprehensive light absorption information of the imaging cross section [24]. The traditional circular scanning mainly irradiates the laser beam from the top to the bottom of the object [25–28], which results in the loss of deeper tissue structure information, while at the same time, the scattering effect of the light cannot effectively illuminate the irregular section and affects the imaging effect. From the perspective of the cost, the array ultrasonic transducer with high performance makes it not only expensive, but also a huge challenge for transducer manufacturing and data acquisition. Once formed, its scanning radius cannot be adjusted according to the size of the imaging targets, which limits the scope and flexibility of its application. There are also some photoacoustic tomography systems that acquire photoacoustic signals point by point through a single-element transducer carried by a rotary motor, whose advantages are their low cost and easy operability. The ultrasonic transducer performs a rotary motion at a predefined speed and completes the acquisition of the photoacoustic signal of the imaging section during the continuous motion process, the continuous scan of data acquisition is faster than a stop-and-go scan [29–31]. In addition, in order to improve the resolution of the photoacoustic imaging system, a negative lens can be mounted on the surface of the ultrasonic transducer to increase the receiving angle [32,33]. Due to the large field-of-view of the virtual point ultrasound transducer, some research has applied virtual point ultrasonic transducers to a photoacoustic imaging system [34–36]. Nie LM et al. developed a virtual point ultrasonic transducer with a large receiving angle and applied it to the photoacoustic tomography system. The signal-to-noise ratio of the system is greatly improved compared with the negative-lens transducers [35]. Hao F. Zhang et al. use a synthetic-aperture focusing technique based on the virtual point detector to obtain a high-resolution photoacoustic image of hemoglobin oxygen saturation in living mice [36].

In this paper, a 3D photoacoustic imaging system based on "dual-element" ultrasonic transducer is proposed under the premise of full-view imaging and uniform illumination. The system contains 1-to-8 fan-out fiber bundles and an optical cage design, which delivers approximately uniform laser radiation toward the imaging phantom and provides an exciting source for completing the photoacoustic tomography of each B-scan (B-scan refers to a cross section of photoacoustic tomography in this paper). The rotary motor carries two customized dual-foci virtual point ultrasonic transducers for photoacoustic data acquisition in the process of rotary motion, while the precise vertical stepper motor drives the imaging target to do vertical movement to complete the 3D scanning of it. The dual-foci virtual point ultrasonic transducer can expand the signal acceptance angle in the short focusing direction because of its large field of view, such that a relatively complete photoacoustic signal can be acquired with fewer detection units, which reduces the cost of the photoacoustic imaging system and accelerates the photoacoustic imaging speed. In the long focusing direction, the narrow sound beam and the annular illumination beam are coupled to form a photoacoustic coplanar mode, which improves the signal to noise ratio of the photoacoustic signal. The pattern of 360° circular laser radiation provides uniform illumination at all angles around the imaging cross section, which facilitates photoacoustic tomography of imaging phantom with irregular surfaces. Furthermore, the feasibility and operability of the 3D-PACT system is validated through systematic testing and 3D photoacoustic experiments on a variety of phantoms.

2. Methods

2.1. Introduction to the 3D-PACT System

Figure 1 shows a schematic of the 3D-PACT system. The system consists of four parts, which are optical illumination, photoacoustic signal acquisition, data processing, and human computer interaction. A pulsed Nd:YAG laser source (Vigour-A-100S, Ziyu, Anshan Liaoning Province, China) with the wavelength of 532 nm was used for the optical illumination. The nanosecond pulsed laser with pulse repetition rate of 20 Hz could provide a laser beam with the pulse width of 5 ns. The laser beam was coupled into the multimode fiber (Ceram Optec, Bonn, Germany, damage threshold was 9.1 mJ/mm^2) located in the fiber coupler after collimating and correcting the optical path. Furthermore, the multimode fiber was divided into eight branches at the output end that were evenly distributed around the water tank. After testing, the coupling ratios of the eight sub-fibers were 9.35%, 9.25%, 9.33%, 9.34%, 9.40%, 9.43%, 9.44%, and 9.41%, respectively, and the total coupling ratio of the fiber was 74.95%. The laser beam out of each branch passed through a convex lens and a plano-convex cylinder lens to form a rectangular strip with a thickness of about 2 mm, as shown in Figure 2a,b, which shows the top view and the front view of the optical path, respectively. The convex lens collimates and corrects the laser beam, while the plano-convex cylinder lens focused the beam in the direction of thickness. Eight laser beams separately passed through the water tank and the transparent quartz bowl, which were placed as shown in Figure 2b, eventually converging in the middle of the tank to form a bright circular area that irradiated the surface of the sample. According to the experimental verification, the light transmittance of the water tank used in this system was about 98%, which was due to the acrylic material that had excellent light transmittance.

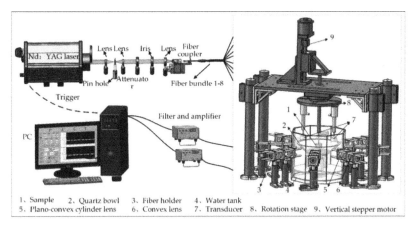

1、Sample   2、Quartz bowl   3、Fiber holder   4、Water tank
5、Plano-convex cylinder lens   6、Convex lens   7、Transducer   8、Rotation stage   9、Vertical stepper motor

Figure 1. Schematic of the 3D-PACT system.

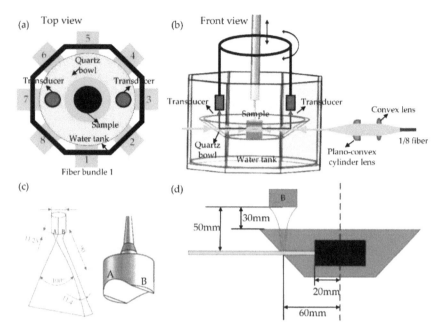

Figure 2. (a) Top view of the core optical path, (b) front view of the core optical path, (c) ultrasonic transducer and its size parameters, and (d) installation method and position diagram of ultrasonic transducer.

It is worth mentioning that this system was the first photoacoustic tomography system for innovatively using the structure of quartz bowl with the characteristics of light transmission and ultrasonic reflection for maintaining the photoacoustic coplanarity of the system while collecting photoacoustic signals. This means that the circular spot was in the same layer as the detection plane of ultrasonic transducer. The slim photoacoustic coplanar configuration eliminated noise from nearby planes, and enables optimal photoacoustic signal excitation and detection. The light transmittance and ultrasonic reflection efficiency are two critical parameters for a quartz bowl, which were 88.25% and 202.465%. The signal generated by the sample was reflected by the quartz bowl and detected by the dual-foci virtual point ultrasonic transducer (Olympus, Tokyo, Japan) placed vertically above the bowl with a 5 MHz central frequency and an 18 mm diameter of the detection plane, as shown in Figure 2c. The surface of the dual-foci transducer adopted the method of dual concave crafting in order to achieve foci in two perpendicular directions, respectively indicated by the letters A and B in Figure 2c. The focal length was 11.25 mm and the directional angle was 100° in direction A, while in direction B, the focal length was about 90.0 mm and the directional angle was 11.4°. It can be concluded from the above parameters that the large receiving angle based on its virtual point in direction A contributed to sparse sampling, while the long focal length in direction B formed a long focal zone (≈52 mm) that determined its applicability for large imaging targets. The installation position of the ultrasonic transducer is shown in Figure 2d. In the 3D-PACT system, two custom-built virtual-point ultrasonic transducers were evenly distributed on the motor rotary table. The rotating motor performed a circular motion according to the preset speed, and the high-speed data acquisition device could complete the collection of the photoacoustic signal in the process of the motor rotation according to the timing trigger signal.

During the whole working process, the photoacoustic signals were obtained through rotating the swivel table 180° driven by a rotary stepping motor and moving the sample 0.1 mm vertically at a time driven by a vertical stepper motor. After amplification (Ultrasonic Transceiver, 5073PR, Olympus, Tokyo, Japan) and digitization (Data Acquisition card, ATS330, AlazarTech, Canada), the raw photoacoustic data was be sent and stored in a personal computer (PC) to reconstruct the 3D photoacoustic image. The detection signal under the large receiving angle held by the virtual point-based ultrasonic transducer used in this system could provide more sample information, which contributed to recovering images with less data. In the reconstruction algorithm, the shape of the sample could be reconstructed by the improved back-projection reconstruction algorithm. The photoacoustic system adopted LabVIEW_2014 to implement the overall control system design, which realized the fully automatic man-machine interaction.

2.2. The Reconstruction Algorithm of the 3D-PACT System

According to the principle of photoacoustic imaging, the homogeneous wave equation can be formulized as:

$$(\nabla^2 - \frac{1}{c^2}\frac{\partial^2}{\partial t^2})p(r\prime,t) = -p_0(r)\frac{d\delta(t)}{dt} \qquad (1)$$

where c is the velocity of sound and p is the initial sound pressure generated by the energy deposition of the pulsed laser. By solving the above wave equation, the photoacoustic signal detected by the ultrasonic transducer at any time t and any position $r\prime$ is obtained, which can be expressed as:

$$p(r\prime,t) = \frac{\partial}{\partial t}[\frac{1}{4\pi c^3 t}\int dr \cdot p_0(r) \cdot \delta(t - \frac{|r\prime - r|}{c})] \qquad (2)$$

The distribution of the original photoacoustic signal can be recovered by using the back-projection reconstruction algorithm [37–39]. Furthermore, the traditional back-projection reconstruction algorithm is shown in Equation (3):

$$p_0(r) = \frac{1}{4\pi c^3}\int dS\frac{1}{t}[\frac{p(r\prime,t)}{t} - \frac{\partial p(r\prime,t)}{\partial t}]|_{t=|r\prime-r|/c} \qquad (3)$$

With the development of the traditional back-projection reconstruction algorithm, a reconstruction algorithm of a ring-shaped array photoacoustic image based on a sensitivity factor is proposed in this study. The photoacoustic signals collected by the transducer were back-projected to each pixel on the arc established by time in the imaging area, which is illustrated in Figure 3a as a schematic diagram. The aforementioned virtual point ultrasonic transducer has a large signal receiving angle (100°, corresponding to the angle Φ in the figure), and the transducer is relatively sensitive to the central region of its receiving scope. Therefore, a variable sensitivity factor $S(\psi)$ was introduced in the algorithm of image reconstruction, and it decreased with the increase of ψ, which is the angle between the connecting line of the pixel and probe and the center line of the detection area. In the process of algorithm implementation, the basic Equation (4) needed to be discretized first:

$$p_0(r) = \sum_{i=1}^{n}[p(r\prime\prime,t) - t\frac{\partial p(r\prime,t)}{\partial t}]|_{t=|r\prime\prime-r|/c} \qquad (4)$$

where $r\prime\prime$ is a distance parameter introduced in the process of discretization. Taking the angle of the receiving signal of the ultrasound probe into account, the back-projection reconstruction algorithm based on sensitivity factor is proposed and can be expressed as:

$$p_0(r) = \sum_{i=1}^{n}S(\psi)_{(r,i)}[p(r\prime\prime,t) - t\frac{\partial p(r\prime,t)}{\partial t}]|_{t=|r\prime\prime-r|/c} \qquad (5)$$

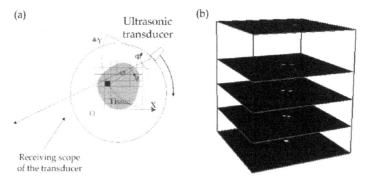

Figure 3. (**a**) A 2D schematic of the improved back-projection reconstruction algorithm, and (**b**) 3D photoacoustic image reconstruction.

In order to perform the 3D image rendering, a large number of B-scans were loaded into Amira 5.4.3 to be superimposed and processed, as shown in Figure 3b.

3. Results

3.1. Performance Test of the 3D-PACT

3.1.1. Feasibility Verification of Quartz Bowl in the System

First of all, the photoacoustic experiments were carried out in the ultrasonic transmission mode and ultrasonic reflection mode to verify the feasibility of the quartz bowl. The schematic diagrams of the two experimental devices are shown in Figure 4a,b, respectively. Black tape was selected as the target due to the strong laser absorption. The energy at the fiber exit was around 100 μJ. The energy E in Table 1 refers to the laser energy of the imaging target surface.

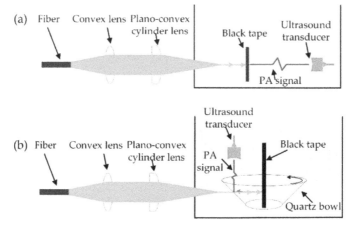

Figure 4. (**a**) Schematic diagram of photoacoustic imaging device in ultrasonic transmission mode. (**b**) Schematic diagram of photoacoustic imaging device in ultrasonic reflection mode.

The ultrasonic reflection efficiency H of the quartz bowl is defined as the ratio of the photoacoustic signal amplitude in the above two signal acceptance modes in the research. The specific expression is:

$$H = G_{ref} / G_{trm} \tag{6}$$

where G_{ref} is the signal amplitude per microjoule laser energy in the ultrasonic reflection mode and G_{trm} is the signal amplitude per microjoule laser energy in the ultrasonic transmission mode. Furthermore, G_{trm} was about 0.01232 V/μJ, found by analyzing the photoacoustic signal obtained from photoacoustic experiment in ultrasonic transmission mode. The quartz bowl was divided into eight areas (Q_1–Q_8), and the quartz bowl's ultrasonic reflection efficiency was tested after subtracting the light attenuation of each area.

Table 1. The ultrasonic reflection efficiency of the quartz bowl.

Areas	E (μJ)	Mean of Signal Amplitude P (V)	G_{ref} = P/E (V/μJ)	H = G_{ref}/G_{trm}
Q_1	88.7614	2.236	0.02519	204.46%
Q_2	88.3206	2.214	0.02507	203.49%
Q_3	88.0611	2.183	0.02480	201.30%
Q_4	88.1574	2.195	0.02490	202.11%
Q_5	88.2163	2.222	0.02520	204.54%
Q_6	88.0742	2.156	0.02448	198.70%
Q_7	88.0688	2.174	0.02469	200.41%
Q_8	88.3672	2.229	0.02522	204.71%

The ultrasonic reflection efficiency of quartz bowl was between 198.70% and 204.71%, which could be seen by noting that the amplitude of the photoacoustic signal was nearly doubled in ultrasonic reflection mode. Experiments showed that the quartz bowl used in the system had a higher efficiency of ultrasonic signal reflection, which was helpful for the collection of photoacoustic signals in PACT.

3.1.2. Resolution Test of the System

In the resolution test experiment, a black sphere with a diameter of 0.3 mm in 2% agar was used as an imaging target and photoacoustic signal source, and was placed in the center of the circular illumination zone, as shown in Figure 5a. Because of its all-black color, the sphere was a good absorber at the laser wavelength of 532 nm used in the system. In the process of data acquisition, the sampling rate was 50 MHz, the number of sampling points in a B-scan was 180 and the total laser energy reaching the surface of the phantom was about 3.2 mJ (the optical fluence of each bundle was 0.7 mJ/mm^2). During the experiment, the imaging target, quartz bowl, and the ultrasonic transducer were all immersed in distilled water to achieve acoustic coupling. The reconstructed photoacoustic image is shown in Figure 5b. It should be noted that the imaging resolution in this research was defined as the full width at half maximum of the normalized absorption intensity at the centerline of photoacoustic image Figure 5b. As shown in Figure 5c, the imaging resolution of the system was 0.31 ± 0.02 mm based on quantitative calculations. In addition, according to the raw data analysis of the photoacoustic signal shown in Figure 5d, it can be seen that the intensity of the photoacoustic signal of the imaging target was relatively uniform at the cross section, and there was no phenomenon showing that the photoacoustic signal was submerged. These also reflected the ability of the system to maintain the photoacoustic coplanar property.

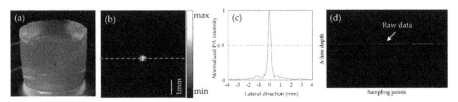

Figure 5. (a) Phantom of resolution test, (b) reconstructed PA image of target, (c) the profile along the dotted line in (b), and (d) raw data of a B-scan.

3.1.3. The Superiority of Photoacoustic Coplanar Configuration and Virtual Point Detection

In the photoacoustic tomography system proposed herein, the narrow sound beam of dual-foci virtual point ultrasonic transducer and the annular illumination beam were coupled to form a photoacoustic coplanar mode based on the quartz bowl with the characteristics of laser transmission and ultrasonic reflection. To prove the advantages of the virtual-point ultrasonic detection and photoacoustic coplanar mode, a comparative experiment was performed using a conventional unfocused ultrasonic transducer (V310, Olympus, Tokyo, Japan). The results of the comparative experiment are shown in Figure 6. In the experiment, the black tape was evenly wrapped around a steel column with a diameter of 25 mm as an imaging phantom, as shown in Figure 6a. The number of sampling points per imaging section was 90, and it took about 9 s to complete a photoacoustic data acquisition of a cross section. Figure 6c,d showed the reconstruction images from the dual-foci virtual point ultrasonic transducer and unfocused ultrasonic transducer, respectively. From a qualitative point of view, it is obvious that the quality of the reconstructed image with the dual-foci transducer was better than its counterpart in terms of qualities such as shape-similarity and image-purity.

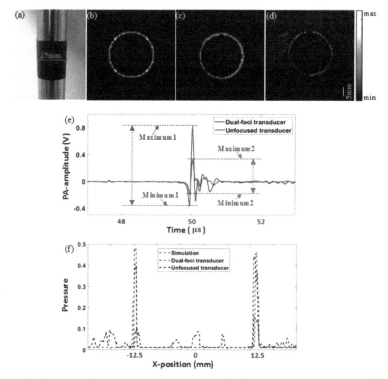

Figure 6. (**a**) Phantom of the comparative experiment. (**b**) Reference standard image by K-Wave simulation. Reconstructed PA images obtained with (**c**) the dual-foci transducer and (**d**) the traditional unfocused transducer. (**e**) A-Line plots from dual-foci transducer and traditional unfocused transducer. (**f**) Absorption intensity distribution at the centerline of (**a**–**c**).

In order to quantitatively analyze the reconstruction quality of Figure 6c,d, the simulation result of the imaging target (shown in Figure 6b) was used as a reference standard. The photoacoustic raw data of Figure 6b was obtained using a K-wave simulation, and then reconstructed using the improved back projection algorithm mentioned above. The performance indicators, such as MSE (mean squared error), PSNR (peak signal to noise ratio), and SSIM (structural similarity) [40–42] were calculated separately according to the following formulae. The MSE index is given as:

$$MSE = \frac{1}{mn}\sum_{i=1}^{m}\sum_{j=1}^{n}\left\|I(i,j) - K(i,j)\right\|^2 \tag{7}$$

The $I(i, j)$ and $K(i, j)$ in the formula represent the reconstructed image and the reference standard image, which refer to the reconstructed photoacoustic image and the K-Wave simulation image in this paper, and the parameters m and n represent the length and width of the image.

PSNR is defined as:

$$PSNR = 20 * \log_{10}(\frac{MAX_1}{\sqrt{MSE}}) \tag{8}$$

where MAX_1 is the gray level of the image. Lastly, the SSIM is given as:

$$SSIM(x,y) = \frac{(2\mu_x\mu_y + c_1)(2\sigma_{xy} + c_2)}{(\mu_x^2 + \mu_y^2 + c_1)(\sigma_x^2 + \sigma_y^2 + c_2)} \tag{9}$$

where x and y represent the two images for comparison, μ_x and μ_y are the mean of the image, σ_x and σ_y represent the variance of image, σ_{xy} is the covariance of x and y, $c_1 = (k_1L)^2$ and $c_2 = (k_2L)^2$ are two constants used to avoid division by zero in the formula, $L = 2^B - 1$ is the range of pixel values, and $k_1 = 0.01, k_2 = 0.03$ are default values.

The results of the above three performance indexes are shown in the following Table 2.

Table 2. Quality evaluation of reconstructed images.

Index	PSNR (dB)	MSE	SSIM
PA image (c) (Dual-foci transducer)	35.9379	0.0051	0.9995
PA image (d) (Unfocused transducer)	15.2504	0.0350	0.9174

It can be seen form Table 2 that the performance indicators SSIM and PSNR of the reconstructed images were greatly improved in the case of virtual-point ultrasound detection, while the value of MSE was obviously reduced. The three performance indicators of the reconstructed photoacoustic image under the dual-foci virtual point transducer were better than those under the unfocused transducer that verified the reliability of the proposed system.

The PSNR of the photoacoustic signal from the dual-foci virtual point ultrasound transducer was 35.9379 dB, which was much larger than that of the conventional unfocused transducer (15.2504 dB), indicating that the dual-foci virtual point transducer had a higher detection sensitivity. Moreover, the high detection sensitivity of the photoacoustic tomography system is further illustrated in Figure 6e, in which the blue and red A-line plots corresponding to the raw photoacoustic signal with the dual-foci virtual-point ultrasonic transducer and the traditional unfocused ultrasonic transducer are depicted, respectively. The two A-line plots were randomly obtained under the premise of the same scanning position and sampling points. It can be seen that the photoacoustic signal from the dual-foci ultrasonic transducer had an amplitude increase of nearly 2.5-fold, which greatly improved the signal to noise ratio of the photoacoustic image. Figure 6f shows the absorption intensity at the position of the centerline of the reconstructed image, and it can be found that the absorption intensity of the center line position of the photoacoustic image obtained by the unfocused ultrasonic transducer was weak (corresponding to the black dotted line in the Figure 6f), and there are signs that the signal was flooded by noise.

In addition, the virtual-point ultrasonic transducer has a large receiving angle, which makes the photoacoustic tomography under sparse sampling possible. In the experiment mentioned above, although the number of sampling points per cross section was 90, the system with virtual-point detection could still perform high quality photoacoustic tomography. However, there were some artifacts in the photoacoustic image under the traditional unfocused transducer. The experiment proved that the system proposed in this paper can improve the imaging speed and reduce the burden of data acquisition device using sparse sampling. The comparative experiments analyzed the photoacoustic signal acquisition mode based on photoacoustic coplanar structure and virtual-point ultrasound detection from a quantitative point of view.

3.1.4. Verification Experiment on Large-Scale Imaging

In order to verify the large-scale imaging characteristics of the system, a photoacoustic tomography experiment was performed using a phantom (a hollow cylinder wrapped with black tape, the diameter is 52 mm), as shown in Figure 7a. Figure 7b shows a photoacoustic image of a cross section. It can be seen that the imaging range of the system could be up to 52 mm in diameter.

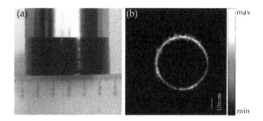

Figure 7. (**a**) Phantom of imaging range test, (**b**) Reconstructed PA image.

To show the 3D photoacoustic imaging in a wide area, four black polyethylene tubes with large spatial distances were designed as imaging targets, as shown in Figure 8a. Figure 8b is a reconstructed photoacoustic image of the scanning section indicated by the dotted line in the real figure. Figure 8c is one of an A-line signal of the B-scan, in which the four signal peaks correspond to the four polyethylene tubes respectively, and the distance between the signal peaks matches the distance between actual tubes. Figure 8d is a 3D photoacoustic image of the imaging target, and the quantized spatial distance is consistent with those shown in Figure 8a,b. It can be seen that the 3D photoacoustic image could more accurately reflect the information of the phantom, such as shape, size, deformation, and so on. According to the photoacoustic images of this test experiment, it can be concluded that even if there was a large space distance, the target shown in Figure 8a could still be clearly imaged.

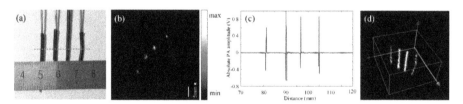

Figure 8. (**a**) Phantom of confirmatory experiment on scope, (**b**) reconstructed PA image of target, (**c**) one of an A-line of (**b**), and (**d**) 3D photoacoustic image.

3.2. Photoacoustic Experiments of Different Kinds of Phantoms

After completing the test experiment of resolution and the verification experiment of a large imaging range, a 3D photoacoustic imaging experiment of the vascular phantom was performed to further verify the feasibility of the imaging system. Two black wires with a diameter of 1 mm were used to make a phantom that simulates blood vessels in tissue, as shown in Figure 9a. In this experiment, photoacoustic signals from 180 positions were acquired on each imaging section, that is, a photoacoustic signal was collected for every 2° of rotation of the ultrasonic probe; it took about 18 s to acquire the information of a B-scan. The step size of the vertical movement of the phantom was 0.2 mm, and 120 B-scans were scanned during the experiment. Figure 9b shows the photoacoustic images of three B-scans and the corresponding raw data. It can be seen that the system can accurately reflect the shape and size of the phantom. Figure 9c shows the three-dimensional photoacoustic image of the phantom, which not only clearly identifies the different parts of the phantom, but also reflects the change of it in the vertical direction, that is, the reconstructed result is highly consistent with the blood vessel phantom. This result indicates that the 3D-PACT is capable of 3D imaging small objects of different spatial distribution and orientation in high quality.

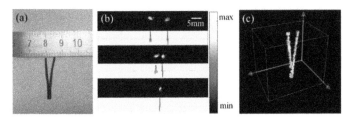

Figure 9. (a) Photograph of a blood vessel phantom, (b) PA images and A-lines of several B-scans, and (c) 3D photoacoustic image.

Subsequently, a 3D photoacoustic experiment was performed on a tumor phantom obtained by soaking the plasticine, whose dimensions are shown in Figure 10a, in black ink. In this experiment, a total of 90 images were obtained by scanning the phantom with a vertical interval of 0.1 mm, and the energy of the pulsed laser reaching the phantom surface was approximately 3.6 mJ. It can be seen that the laser could penetrate the imaging phantom and the laser absorption characteristics of each B-scan, and each point on it could be well-reflected from both two-dimensional and three-dimensional photoacoustic images demonstrated in Figure 10b,c separately. By comparing Figure 10a,c, it can be seen that the size of the absorption region on the photoacoustic image was substantially the same as the actual size of the corresponding section, which verified the feasibility and reliability of this system by showing that the shape and size of tumor analogs can be well-reflected through 3D photoacoustic imaging.

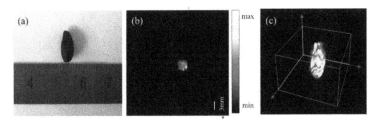

Figure 10. (a) Photograph of a tumor phantom, (b) PA image of a B-scan, and (c) three-dimensional photoacoustic image.

Finally, in order to verify the photoacoustic imaging of some irregularly shaped tissues, 3D photoacoustic experiments were carried out on complex-shaped phantoms. For example, a phantom made of knotted black wire with a diameter of about 0.6 mm is shown in Figure 11a. The sampling frequency, sampling position, laser energy, and z-axis spacing of this experiment were the same as those above. Figure 11b is a 3D photoacoustic image reconstructed by using the detected photoacoustic signals detected from 90 B-scans spaced 0.1 mm apart, indicating that the reconstructed photoacoustic image could present the actual shape and dimensions of the target clearly and with high quality.

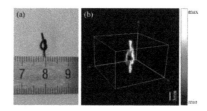

Figure 11. (**a**) Photograph of a knot phantom and (**b**) 3D photoacoustic image.

4. Discussion and Conclusions

A 3D-PACT imaging system based on full-view illumination and detection was proposed and developed in this study. In the design of optical path, the system adopted a mode of circular illumination with the even arrangement of eight laser beams around the water tank to ensure that the imaging target received more uniform laser radiation, which improved the low-efficiency of photoacoustic imaging whose illumination was from the top to the bottom. Second, drawing on the design of photoacoustic coaxial confocal in photoacoustic microscopy imaging technology, a photoacoustic coplanar structure was proposed by means of a quartz bowl with light-transmitting and ultrasound-reflecting properties. The system extended the photoacoustic coupling from a single dimension to a two-dimensional space, increasing the photoacoustic detection range of the imaging section and improving the signal to noise ratio of the photoacoustic signal. In addition, the dual-foci virtual point ultrasonic transducer with a large receiving angle improved the imaging speed of the system via sparse sampling. In the reconstruction method of the photoacoustic image, taking the receiving angle of ultrasound transducer into account, the improved back-projection reconstruction algorithm based on sensitivity factor was achieved, and the photoacoustic image reconstruction using the algorithm took 1.4287 s.

The practicality of the illumination method and the designed structure enable the system to perform clear and high-quality 3D photoacoustic imaging on targets with complex surfaces or requiring a large imaging range. The feasibility of the system was verified through systematic testing of resolution, imaging range, and photoacoustic coplanar property and 3D photoacoustic experiments on various phantoms. The best compromise was reached between imaging performance and system cost.

The imaging speed of the system was about 9 s in one cross section (90 sampling points per B-scan, it took about 15 min to scan 100 B-scans), and had not reached the requirements of fast imaging; however, the phantom experiments conducted in this study have demonstrated the characteristics of the system. The next step will be focused on the research of sparse circular array ultrasonic transducer based on virtual point and resolution improvement. In practical applications, the system is expected to be used for corresponding work in the pathological study of an in vitro tumor and the research of targeted photoacoustic probes for breast cancer cells. Moreover, the 3D reconstruction will be further investigated.

Author Contributions: Design and construction of the 3D-PACT, M.S., D.H., and L.M.; Completion of photoacoustic imaging experiments, D.H., W.Z., and Y.Q.; Control system design, M.S., D.H., and L.M.; System debugging and improvement, Y.L., Y.Q., and W.Z.; Algorithm implementation, M.S., D.H., and Y.L.; Paper writing, D.H., W.Z., and M.S.

Funding: This research was funded by the National Key R & D Program of China (Grant No. 2017YFE0121000 and No. 2018YFC0114800), National Natural Science Foundation of China (Grant No. 11574064 and No. 11874133), Shandong Provincial Natural Science Foundation, China (Grant No. ZR2017MF041 and ZR2018MF026), the Science and Technology Development Plan Project of Shandong Province (Grant No. 2018GGX103047 and 2016GGX103032), and the Development Plan of Chinese Academy of Sciences and Wego Group (Grant No. 2017011).

Acknowledgments: This work was partly supported by the Institute of Biomedical and Health Engineering at Shen Zhen Institute of Advanced Technology, Chinese Academy of Sciences (SIAT).

Conflicts of Interest: The authors declare no conflict of interest.

References

1. Xu, M.; Wang, L.V. Photoacoustic imaging in biomedicine. *Rev. Sci. Instrum.* **2006**, *77*, 41101. [CrossRef]
2. Wang, L.V.; Hu, S. Photoacoustic Tomography: In Vivo Imaging from Organelles to Organs. *Science* **2012**, *335*, 1458–1462. [CrossRef]
3. Beard, P. Biomedical photoacoustic imaging. *Interface Focus* **2011**, *1*, 602. [CrossRef]
4. Wang, L.V.; Wu, H. Biomedical optics: Principles and imaging. *J. Biomed. Opt.* **2008**, *13*, 049902. [CrossRef]
5. Liu, T.; Sun, M.; Feng, N.; Wu, Z.; Shen, Y. Multiscale Hessian filter-based segmentation and quantification method for photoacoustic microangiography. *Chin. Opt. Lett.* **2015**, *13*, 091701.
6. Lin, X.; Sun, M.; Feng, N.; Hu, D.; Shen, Y. Monte Carlo light transport-based blood vessel quantification using linear array photoacoustic tomography. *Chin. Opt. Lett.* **2017**, *15*, 111701.
7. Wang, X.; Pang, Y.; Ku, G.; Xie, X.; Stoica, G.; Wang, L.V. Noninvasive laser-induced photoacoustic tomography for structural and functional in vivo imaging of the brain. *Nat. Biotechnol.* **2003**, *21*, 803–806. [CrossRef]
8. Wang, L.V.; Yao, J. A practical guide to photoacoustic tomography in the life sciences. *Nat. Methods* **2016**, *13*, 627–638. [CrossRef]
9. Kruger, R.A.; Lam, R.B.; Reinecke, D.R.; Rio, S.P. Photoacoustic angiography of the breast. *Med. Phys.* **2010**, *37*, 6096–6100. [CrossRef]
10. Wang, S.; Lin, J.; Wang, T.; Chen, X.; Huang, P. Recent Advances in Photoacoustic Imaging for Deep-Tissue Biomedical Applications. *Theranostics* **2016**, *6*, 2394–2413. [CrossRef]
11. Wang, H.; Liu, C.; Gong, X.; Hu, D.; Lin, R.; Sheng, Z.; Zheng, C.; Yan, M.; Chen, J.; Cai, L.; et al. In vivo photoacoustic molecular imaging of breast carcinoma with folate receptor-targeted indocyanine green nanoprobes. *Nanoscale* **2014**, *6*, 14270–14279. [CrossRef]
12. Cao, M.; Yuan, J.; Du, S.; Xu, G.; Wang, X.; PL, C.; Liu, X. Full-view photoacoustic tomography using asymmetric distributed sensors optimized with compressed sensing method. *Biomed. Signal Process. Control* **2015**, *21*, 19–25. [CrossRef]
13. Wang, X.; Roberts, W.W.; Carson, P.L.; Wood, D.P.; Fowlkes, J.B. Photoacoustic tomography: A potential new tool for prostate cancer. *Biomed. Opt. Express* **2010**, *1*, 1117. [CrossRef]
14. Wurzinger, G.; Nuster, R.; Schmitner, N.; Gratt, S.; Meyer, D.; Paltauf, G. Simultaneous three-dimensional photoacoustic and laser-ultrasound tomography. *Biomed. Opt. Express* **2013**, *4*, 1380. [CrossRef]
15. Xie, Z.; Tian, C.; Chen, S.L.; Ling, T. 3D high resolution photoacoustic imaging based on pure optical photoacoustic microscopy with microring resonator. In *Proceedings of the SPIE 8943, Photons Plus Ultrasound: Imaging and Sensing 2014*; International Society for Optics and Photonics: Bellingham, WA, USA, 2014.
16. Moradi, H.; Honarvar, M.; Tang, S.; Salcudean, S.E. Iterative photoacoustic image reconstruction for three-dimensional imaging by conventional linear-array detection with sparsity regularization. In *Proceedings Volume 10064, Photons Plus Ultrasound: Imaging and Sensing*; International Society for Optics and Photonics: Bellingham, WA, USA, 2017.
17. Tan, Y.; Xia, K.; Ren, Q.; Li, C. Three-dimensional photoacoustic imaging via scanning a one dimensional linear unfocused ultrasound array. *Opt. Express* **2017**, *25*, 8022–8028. [CrossRef]
18. Vaithilingam, S.; Ma, T.J.; Furukawa, Y.; Wygant, I.O.; Zhuang, X.; De La Zerda, A.; Oralkan, O.; Kamaya, Y.; Gambhir, S.S.; Jeffrey, R.B.; et al. Three-dimensional photoacoustic imaging using a two-dimensional CMUT array. *IEEE Trans. Ultrason. Ferroelectr. Freq. Control* **2009**, *56*, 2411–2419. [CrossRef]
19. Paltauf, G.; Nuster, R.; Burgholzer, P.; Haltmeier, M. Three-dimensional photoacoustic tomography using acoustic line detectors. *Proc. SPIE* **2007**, *6437*, 64370N.
20. Ermolayev, V.; Dean-Ben, X.L.; Mandal, S.; Ntziachristos, V.; Razansky, D. Simultaneous visualization of tumour oxygenation, neovascularization and contrast agent perfusion by real-time three-dimensional optoacoustic tomography. *Eur. Radiol.* **2016**, *26*, 1843–1851. [CrossRef]
21. Wang, B.; Xiang, L.; Jiang, M.S.; Yang, J.; Zhang, Q.; Carney, P.R.; Jiang, H. Photoacoustic tomography system for noninvasive real-time three-dimensional imaging of epilepsy. *Biomed. Opt. Express* **2012**, *3*, 1427–1432. [CrossRef]
22. Hui, L.; Yu, C.; Hongbo, L.; Dong, P.; Yukun, Z.; Kun, W.; Jie, T. High-speed and 128-channel multi-spectral photoacoustic tomography system for small animal. *Laser Technol.* **2017**, *41*, 669–674.
23. Razansky, D.; Buehler, A.; Ntziachristos, V. Volumetric real-time multispectral optoacoustic tomography of biomarkers. *Nat. Protoc.* **2011**, *6*, 1121–1129. [CrossRef]

24. Xia, J.; Chatni, M.R.; Maslov, K.; Guo, Z.; Wang, K.; Anastasio, M.; Wang, L.V. Whole-body ring-shaped confocal photoacoustic computed tomography of small animals in vivo. *J. Biomed. Opt.* **2012**, *17*, 050506. [CrossRef]

25. Wong, T.T.; Zhang, R.; Zhang, C.; Hsu, H.C.; Maslov, K.I.; Wang, L.; Shi, J.; Chen, R.; Shung, K.K.; Zhou, Q.; et al. Label-free automated three-dimensional imaging of whole organs by microtomy-assisted photoacoustic microscopy. *Nat. Commun.* **2017**, *8*, 1386. [CrossRef]

26. Paltauf, G.; Hartmair, P.; Kovachev, G.; Kovachec, G.; Nuster, R. Piezoelectric line detector array for photoacoustic tomography. *Photoacoustics* **2017**, *8*, 28–36. [CrossRef]

27. Li, C.; Aguirre, A.; Gamelin, J.; Maurudis, A.; Zhou, Q.; Wang, L.V. Real-time photoacoustic tomography of cortical hemodynamics in small animals. *J. Biomed. Opt.* **2010**, *15*, 010509. [CrossRef]

28. Xia, J.; Chatni, M.; Maslov, K.; Wang, L.V. Anatomical and metabolic small-animal whole-body imaging using ring-shaped confocal photoacoustic computed tomography. In *Proceedings of the Photons Plus Ultrasound: Imaging & Sensing*; International Society for Optics and Photonics: Bellingham, WA, USA, 2013.

29. Ku, G.; Wang, X.; Stoica, G.; Wang, L.V. Multiple-bandwidth photoacoustic tomography. *Phys. Med. Biol.* **2004**, *49*, 1329–1338. [CrossRef]

30. Kruger, R.A.; Liu, P.Y.; Fang, Y.R.; Appledorn, C.R. Photoacoustic ultrasound (PAUS)—Reconstruction tomography. *Med. Phys.* **1995**, *22*, 1605–1609. [CrossRef]

31. Ma, R.; Taruttis, A.; Ntziachristos, V.; Razansky, D. Multispectral optoacoustic tomography (MSOT) scanner for whole-body small animal imaging. *Opt. Express* **2009**, *17*, 21414–21426. [CrossRef]

32. Pramanik, M.; Ku, G.; Wang, L.V. Tangential resolution improvement in thermoacoustic and photoacoustic tomography using a negative acoustic lens. *J. Biomed. Opt.* **2009**, *14*, 024028. [CrossRef]

33. Li, C.; Ku, G.; Wang, L.V. Negative lens concept for photoacoustic tomography. *Phys. Rev. E* **2008**, *78*, 021901. [CrossRef]

34. Perekatova, V.V.; Kirillin, M.Y.; Turchin, I.V.; Subochev, P.V. Combination of virtual point detector concept and fluence compensation in acoustic resolution photoacoustic microscopy. In Proceedings of the the 6th International Symposium on Topical Problems of Biophotonics (TPB), St Petersburg, Russia, 28 July–3 August 2017.

35. Nie, L.; Guo, Z.; Wang, L.V. Photoacoustic tomography of monkey brain using virtual point ultrasonic transducers. *J. Biomed. Opt.* **2011**, *16*, 076005. [CrossRef]

36. Zhang, H.F.; Menglin, L.; Konstantin, M.; Stocia, G.; Wang, L.V. Three-dimensional photoacoustic imaging of subcutaneous microvasculature in vivo. *Proc. SPIE* **2006**, *6086*, 60861E.

37. Minghua, X.; Wang, L.V. Universal back-projection algorithm for photoacoustic computed tomography. *Phys. Rev. E Stat. Nonlinear Soft Matter Phys.* **2005**, *71* 1 Pt 2, 016706.

38. Xu, M.; Wang, L.V. Pulsed-microwave-induced thermoacoustic tomography: Filtered backprojection in a circular measurement configuration. *Med. Phys.* **2002**, *29*, 1661–1669. [CrossRef] [PubMed]

39. Duoduo, H.; Zheng, S.; Yuan, Y. Reconstruction of Intravascular Photoacoustic Images Based on Filtered Back-projection Algorithm. *Chin. J. Biomed. Eng.* **2016**, *35*, 9–19.

40. Wang, Z.; Simoncelli, E.P.; Bovik, A.C. Multi-scale structural similarity for image quality assessment. In Proceedings of the IEEE Asilomar Conference on Signals, Pacific Grove, CA, USA, 9–12 November 2003; Volume 2, pp. 1398–1402.

41. Mandal, S.; Sudarshan, V.P.; Nagaraj, Y.; Dean-Ben, X.L.; Razansky, D. Multiscale edge detection and parametric shape modeling for boundary delineation in optoacoustic images. In Proceedings of the 2015 37th Annual International Conference of the IEEE Engineering in Medicine and Biology Society (EMBC) 2015, Milan, Italy, 25–29 August 2015; pp. 707–710.

42. Kazakeviciute, A.; Ho CJ, H.; Olivo, M. Multispectral Photoacoustic Imaging Artifact Removal and Denoising Using Time Series Model-Based Spectral Noise Estimation. *IEEE Trans. Med. Imaging* **2016**, *35*, 2151–2163. [CrossRef]

Article

Full Field Inversion in Photoacoustic Tomography with Variable Sound Speed

Gerhard Zangerl [1], Markus Haltmeier [1,*], Linh V. Nguyen [2,3] and Robert Nuster [4]

[1]   Department of Mathematics, University of Innsbruck Technikerstraße 13, 6020 Innsbruck, Austria;
      gerhard.zangerl@uibk.ac.at
[2]   Department of Mathematics, University of Idaho 875 Perimeter Dr, Moscow, ID 83844, USA;
      lnguyen@uidaho.edu
[3]   Faculty of Information Technology, Industrial University of Ho Chi Minh City, Ho Chi Minh 71406, Vietnam
[4]   Department of Physics, University of Graz, Universitätsplatz 5, 8010 Graz, Austria; ro.nuster@uni-graz.at
*   Correspondence: markus.haltmeier@uibk.ac.at

Received: 23 January 2019; Accepted: 8 April 2019; Published: 15 April 2019

Abstract: To accelerate photoacoustic data acquisition, in [R. Nuster, G. Zangerl, M. Haltmeier, G. Paltauf (2010). Full field detection in photoacoustic tomography. Optics express, 18(6), 6288–6299] a novel measurement and reconstruction approach has been proposed, where the measured data consist of projections of the full 3D acoustic pressure distribution at a certain time instant T. Existing reconstruction algorithms for this kind of setup assume a constant speed of sound. This assumption is not always met in practice and thus can lead to erroneous reconstructions. In this paper, we present a two-step reconstruction method for full field detection photoacoustic tomography that takes variable speed of sound into account. In the first step, by applying the inverse Radon transform, the pressure distribution at the measurement time is reconstructed point-wise from the projection data. In the second step, a final time wave inversion problem is solved where the initial pressure distribution is recovered from the known pressure distribution at time T. We derive an iterative solution approach for the final time wave inversion problem and compute the required adjoint operator. Moreover, as the main result of this paper, we derive its uniqueness and stability. Our numerical results demonstrate that the proposed reconstruction scheme is fast and stable, and that ignoring sound speed variations significantly degrades the reconstruction.

Keywords: photoacoustic tomography; full-field detection; wave equation; final time inversion; uniqueness; stability; iterative reconstruction

1. Introduction

Photoacoustic tomography (PAT) is a hybrid imaging modality that combines high spatial resolution of ultrasound and high contrast of optical tomography [1–5]. In PAT, a semitransparent sample is illuminated by a short laser pulse. As a result, parts of the optical energy are absorbed inside the sample. This causes an initial pressure distribution and a subsequent acoustic pressure wave. The pressure wave is detected outside the investigated object and used to recover an image of the interior. In standard PAT, the induced pressure is measured on a detection surface as a function of time. In the case of constant sound speed and when the observation surface exhibits a special geometry (planar, cylindrical, spherical), the initial pressure distribution can be recovered by closed-form inversion formulas; see [6–20] and references therein. While some algorithms take the size and other properties of the detectors into account [21–24], most algorithms assume that the acoustic pressure is known point-wise on a detection surface for all times in the measurement interval. Due to the finite width of the commonly used piezoelectric elements this assumption is only approximately satisfied. One approach to address this issue uses the concept of integrating detectors [25]. Integrating detectors measure integrals of the acoustic pressure over planes, lines or circles and several inversion methods have been derived in the literature [26–29].

Inspired by the concept of integrating line detectors, a full field detection and reconstruction method that uses projections of the full acoustic field surrounding the sample has been developed in [30,31]. The pressure field within the depth of field is imaged by the use of an optical phase contrast imaging system with a CCD camera. In this way, one obtains 2D projections of the pressure field at a time instant T. Reconstructing 2D projections of the initial pressure has been investigated in [32,33]. In [30,31] we go one step further and show that projection data from different directions allow for a full 3D reconstruction of the initial pressure by Radon or Fourier transform techniques. We refer to this approach as full field detection photoacoustic tomography (FFD-PAT). A photograph of the FFD-PAT setup developed in Graz is shown in Figure 1. For practical aspects and a detailed description of the working principle of the FFD-PAT detection system we refer to the original works [30,31]. Existing image reconstruction methods for FFD-PAT data assume a constant speed of sound. However, there are relevant cases when the assumption of constant speed of sound is inaccurate [34,35]. For example, it is known that acoustic properties vary within female human breasts. Consequently, for accurate image reconstruction, variable speed of sound has to be incorporated in the wave propagation model. Iterative methods are capable to deal with this assumption. In the case of standard PAT, such methods have been studied in [36–41] for spatially variable speed of sound, that is smooth and bounded from below. Moreover, it is assumed to satisfy the so-called non-trapping condition, which means that the supremum of the lengths of all geodesics connecting any two points inside the volume enclosed by the measurement surface S is finite. Under these assumptions, it is known that the initial pressure can be stably reconstructed from pressure data given on $S \times [0, T]$.

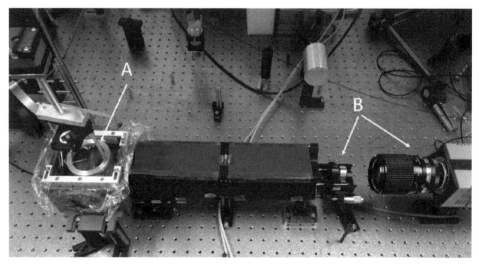

Figure 1. FFD-PAT SETUP DEVELOPED IN GRAZ. The sample to be imaged is mounted and rotated on stage A. Full field projections of $p(\cdot, T)$ are recorded with the CCD camera B using an optical phase contrast technique. For a detailed description of the working principle, see the original works [30,31].

In this paper, for the first time, we study image reconstruction in FFD-PAT with a spatially variable speed of sound. We will give a precise mathematical formulation of FFD-PAT and describe the inverse problem we are dealing with (see Section 2). For its solution, we propose a two-step process. In the first step, the acoustic pressure at time T is reconstructed pointwise from the FFD data. In the second step, we recover the desired initial pressure distribution from the known valued of the pressure at time T. The first step can be approximated by inverting the well-known Radon transform. The second step consists in a final time wave inversion problem with spatially varying speed of sound. To the best of our knowledge, the latter has not been addressed in the literature so far. We develop iterative reconstruction methods based on an explicit computation of an adjoint problem. As main theoretical results of this paper, we establish uniqueness and stability of the final time wave inversion problem (see Section 3). In particular, this implies linear convergence for the proposed iterative reconstruction methods. In Section 4, we present numerical results demonstrating that the proposed algorithm is efficient and stable. Moreover, the presented numerical results clearly highlight the importance of taking sound speed variations into account in FFD-PAT image reconstruction.

2. Full Field Detection Photoacoustic Tomography

In this section, we describe a mathematical model for FFD-PAT including the variable sound speed case, and state the inverse problem under consideration. Additionally, we outline the proposed two-step reconstruction procedure and formulate the final time inverse problem.

2.1. Mathematical Model

In the case of variable sound speed, acoustic wave propagation in PAT is commonly described by the initial value problem [34,38,39,41]

$$p_{tt}(\mathbf{x}, t) - c^2(\mathbf{x})\Delta p(\mathbf{x}, t) = 0, \qquad (\mathbf{x}, t) \in \mathbb{R}^3 \times \mathbb{R}_{>0} \qquad (1)$$

$$p(\mathbf{x}, 0) = f(\mathbf{x}), \qquad \mathbf{x} \in \mathbb{R}^3 \qquad (2)$$

$$p_t(\mathbf{x}, 0) = 0, \qquad \mathbf{x} \in \mathbb{R}^3. \qquad (3)$$

here $c(\mathbf{x}) > 0$ is the sound speed at location $\mathbf{x} \in \mathbb{R}^3$ and $f \in C_0^\infty(\mathbb{R}^3)$ is the initial pressure distribution that encodes the inner structure of the object. It is assumed that the object is contained inside the ball $B(0, R) = \{\mathbf{x} \in \mathbb{R}^3 \mid \|\mathbf{x}\| < R\}$, of radius R centered at the origin, and that the sound speed is smooth, positive and has the constant value c_0 outside $B(0, R)$. We denote by $C_0^\infty(B(0, R))$ the set of all smooth functions $f\colon \mathbb{R}^3 \to \mathbb{R}$ that vanish outside $B(0, R)$.

In FFD-PAT, linear projections (integrals along straight lines) of the 3D pressure field $p(\cdot, T)$ for a fixed time $T > 0$ are recorded. As illustrated in Figure 1, this can be implemented using a special phase contrast method and a CCD-camera that records full field projections of the pressure field [30,31]. The linear projections are collected for rotation angles $\theta \in [0, \pi]$ around the $e_3 = (0, 0, 1)$ axis and are given by

$$g_R(\theta, s, z) = \int_{\mathbb{R}} p(s\cos(\theta) - t\sin(\theta), s\sin(\theta) + t\cos(\theta), z, T) \, dt, \qquad (4)$$

for $(\theta, s, z) \in M_R := \{(\theta, s, z) \in [0, \pi] \times \mathbb{R}^2 \mid s^2 + z^2 \geq R^2\}$. Here M_R determines the set of admissible projections and the defining condition $s^2 + z^2 \geq R^2$ means that in practice only pressure integrals over those lines are recorded, which do not intersect the possible support $B(0, R)$ of the imaged object; compare Figure 2.

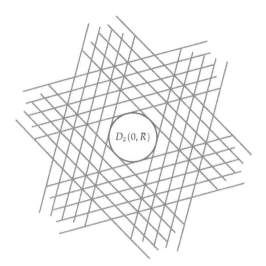

Figure 2. HORIZONTAL CROSS-SECTION OF THE FFD-PAT SETUP.. Full field projections are measured over lines that do not intersect the ball $B(0, R)$. Consequently, for any plane $\mathbb{R}^2 \times \{z\}$, integrals of $p(\,\cdot\,, T)$ over those lines are given that do not intersect the disc $D_z(0, R) := B(0, R) \cap (\mathbb{R}^2 \times \{z\})$. For planes with $|z| \leq R$, this yields the exterior problem for the Radon transform.

2.2. Description of the Inverse Problem

In order to describe the inverse problem of FFD-PAT in a compact way, we introduce some further notation. First, we define the operator

$$\mathbf{W}_T \colon C_0^\infty(B(0, R)) \to C_0^\infty(\mathbb{R}^3) \colon f \mapsto p(\,\cdot\,, T)\,, \tag{5}$$

where p denotes the solution of (1)–(3). The operator $\mathbf{W}_T$ maps the initial data f to the solution (full field) of the wave Equations (1)–(3) at the given measurement time $T > 0$. Second, we define the X-ray transform

$$\mathbf{X} \colon C_0^\infty\left(\mathbb{R}^3\right) \to C_0^\infty\left([0, \pi] \times \mathbb{R}^2\right)$$
$$(\mathbf{X}h)(\theta, s, z) := \int_{\mathbb{R}} h(s\cos(\theta) - t\sin(\theta), s\sin(\theta) + t\cos(\theta), z)\, dt \tag{6}$$

Note that for any fixed $z \in \mathbb{R}$, the function $(\mathbf{X}h)(\,\cdot\,, z)$ is the Radon transform of $h(\,\cdot\,, z)$ in the horizontal plane $\mathbb{R}^2 \times \{z\}$. Finally, we define the restricted X-ray transform

$$\mathbf{X}_R \colon C_0^\infty\left(\mathbb{R}^3\right) \to C^\infty\left(M_R\right) \colon h \mapsto (\mathbf{X}h)|_{M_R}\,. \tag{7}$$

For $|z| \leq R$, $(\mathbf{X}_R h)(\,\cdot\,, z)$ is the exterior Radon transform of $h(\,\cdot\,, z)$ and consist of integrals over lines not intersecting the disc $\{(x, y) \in \mathbb{R}^2 \mid x^2 + y^2 < R^2 - z^2\}$; compare Figure 2. Otherwise, $(\mathbf{X}_R h)(\,\cdot\,, z)$ coincides with the standard Radon transform of $h(\,\cdot\,, z)$.

Using the operator notation introduced above we can write the inverse problem of FFD-PAT in the form

$$\text{Recover } f \text{ from data} \quad g_R = \mathbf{X}_R \mathbf{W}_T f + \text{noise}\,. \tag{8}$$

Evaluation of $\mathbf{X}_R \mathbf{W}_T f$ is referred to as the forward problem in FFD-PAT. In this paper, we study the solution of the inverse problem (8).

2.3. Two-Step Reconstruction

One possible approach to solve the inverse problem of FFD-PAT is to directly recover f from data in (8) via iterative methods. Typically, each iteration step requires the evaluation of the forward operator $\mathbf{X}_R \mathbf{W}_T$ and its adjoint $(\mathbf{X}_R \mathbf{W}_T)^* = \mathbf{W}_T^* \mathbf{X}_R^*$. In this paper, we consider a two-step approach where we first invert $\mathbf{X}_R$ via a direct method and then use an iterative method to invert $\mathbf{W}_T$. This avoids repeated and time consuming evaluation of $\mathbf{X}_R$ and its adjoint. The proposed reconstruction method consists of the following two steps.

- INVERSE RADON TRANSFORM: Assume that projection data $g_R = \mathbf{X}_R \mathbf{W}_T f$ are given and consider the zero extension $g: [0, \pi] \times \mathbb{R}^2 \to \mathbb{R}$ by $g(\theta, x, z) = g_R(\theta, x, z)$ for $(\theta, x, z) \in M_R$ and $g(\theta, x, z) = 0$ otherwise. We then define an approximation to $\mathbf{W}_T f$ by applying an inversion formula of the Radon transform in planes $\mathbb{R}^2 \times \{z\}$. Using the well-known filtered backprojection inversion formula for the Radon transform [42] yields

$$\mathbf{W}_T f(x, y, z) \simeq \mathbf{X}^\sharp g(x, y, z) := \frac{1}{2\pi^2} \int_0^\pi \text{P.V.} \int_{\mathbb{R}} \frac{(\partial_s g)(\theta, s, z) \, ds}{(x \cos(\theta) + y \sin(\theta)) - s} d\theta,$$

  where P.V. denotes the principal value integral.
- FINAL TIME WAVE INVERSION: For the second step, assume that an approximation $h \simeq \mathbf{W}_T f$ to the 3D acoustic field at time T is given. Recovering the initial pressure then yields the final time wave inversion problem

$$\text{Recover } f \text{ from data} \quad h = \mathbf{W}_T f + \text{noise}. \tag{9}$$

To the best of our knowledge, the final time wave inversion problem (9) has not been considered so far and its investigation will be the main theoretical focus of this work.

For solving the wave inversion problem (9), we propose iterative solution methods that are described in Section 3. As main theoretical results, we derive its uniqueness and stability.

Another possible approach for solving the first step would be to work with the exterior Radon transform [43–45]. Instead, we invert the standard Radon transform after setting the missing values of $\mathbf{X}\mathbf{W}_T f$ to zero. We observe numerically that the missing values are indeed close to zero for large enough measurement time T. Although we do not have a rigorous mathematical proof for this claim, it is supported by at least two facts. First, in the case of a non-trapping sound speed, the known decay estimate [46] for the solution p of (1)–(3) with initial data f supported in $B(0, R)$ states

$$\left| \frac{\partial^{k+|m|} p(\mathbf{x}, t)}{\partial_t^k \partial_{\mathbf{x}}^{|m|}} \right| \leq \gamma e^{-\beta t} \|f\|_2 \quad \text{for } (\mathbf{x}, t) \in B(0, R) \times (T, \infty) \tag{10}$$

for all $(k, m) \in \mathbb{N}^2$. Here $\beta > 0$ is a constant only depending on $c(\mathbf{x})$ and T, and γ is a constant depending on $B(0, R)$. Second, in the case of constant sound speed, the X-ray transform reduces (1)–(3) to a 2D wave equation with initial data $\mathbf{X}f$ supported in a disc of radius R. Thus, in the constant sound speed case, $\mathbf{X}(p(\cdot, t))$ describes 2D propagation which rapidly decays inside the disc. Formulating and proving precise statements in such a direction, however, is an interesting unsolved problem.

3. Final Time Wave Inversion Problem for Variable Sound Speed

In this section we study the final time wave inversion problem (9), where the forward operator $\mathbf{W}_T: f \mapsto p(\cdot, T)$ is defined by (5). According to standard results for the wave equation [47], the forward operator extends to a bounded linear operator $\mathbf{W}_T: L^2(B(0, R)) \to L^2(\mathbb{R}^3)$. Below we establish uniqueness and stability results and derive an iterative reconstruction algorithm. Note that for constant sound speed, recovering the function f from the solution at time T of (1) with initial data $(0, f)$ instead of $(f, 0)$, is equivalent to the inversion from spherical means with fixed radius.

Uniqueness and an inversion method for this problem has been obtained in the classical book of Fritz John [48]. However, neither for the case of initial data $(f, 0)$ nor in the variable sound speed case we are aware of any similar results.

3.1. Uniqueness and Stability Theorem

The following theorem is the main theoretical result of this paper and states that the final time wave inversion problem (9) has a unique solution that stably depends on the right-hand side.

Theorem 1 (Uniqueness and stability of inverting $\mathbf{W}_T$). *For a non-trapping speed of sound, the operator* $\mathbf{W}_T \colon L^2(B(0, R)) \to L^2(\mathbb{R}^3)$ *is injective and bounded from below.*

The proof of Theorem 1 is presented in the following Section 3.2. See [19,41,49] for related results and methods in the case of standard PAT data. Theorem 1 states that

$$ b := \inf \left\{ \frac{\|\mathbf{W}_T f\|_{L^2(\mathbb{R}^3)}}{\|f\|_{L^2(B(0,R))}} \,\middle|\, f \in L^2(B(0, R)) \right\} > 0. \tag{11} $$

In particular, $\mathbf{W}_T \colon L^2(B(0, R)) \to \mathrm{Ran}(\mathbf{W}_T)$ has a bounded inverse, where $\mathrm{Ran}(\mathbf{W}_T)$ denotes the range of $\mathbf{W}_T$. The latter implies that gradient based iterative methods for (9) converge linearly, similar to the case of standard PAT data; see [38].

3.2. Proof of Theorem 1

Let c be a non-trapping sound speed. We first prove two crucial Lemmas, from which Theorem 1 follows rather quickly.

Lemma 1. *Assume that* $\mathbf{W}_T f = 0$, *then* $f = 0$. *That is,* $\mathbf{W}_T$ *is injective.*

Proof. Let p denote the solution of (1)–(3). We will construct a solution $\bar{p}$ of the wave equation which is periodic in time with period $4T$ such that $\bar{p} = p$ on $\mathbb{R}^3 \times [0, T]$. Once this is done, we obtain $f = \bar{p}(\cdot, 0) = \bar{p}(\cdot, 4^n T)$ for any n. Using (10), we arrive at

$$ \forall \mathbf{x} \in \mathbb{R}^3 \colon \quad f(\mathbf{x}) = \lim_{n \to \infty} \bar{p}(\mathbf{x}, 4^n T) = 0. $$

It remains to construct the above-mentioned solution $\bar{p}$ of the wave equation. The idea is to properly reflect the solution p in the time variable t through the time moments $t = T, 2T, \ldots$, as follows. We first construct $\bar{p}$ on $[0, 2T]$ by the odd reflection of p through the moment $t = T$: $\bar{p}(\cdot, T) = p(\cdot, T)$ for $t \in [0, T]$ and $\bar{p}(\cdot, T) = -p(\cdot, 2T - t)$ for all $t \in [T, 2T]$. Since $p(\cdot, T) = 0$ on $\mathbb{R}^3$, we obtain that $\bar{p}$ and $\bar{p}_t$ are continuous at $t = T$. Therefore, p is continuous on $[0, 2T]$ and solves the wave equation on that interval. Next note that $\bar{p}_t(\cdot, 2T) = -\bar{p}_t(\cdot, 0) = 0$ on $\mathbb{R}^3$. By the even reflection through $t = 2T$: $\bar{p}(\cdot, T) = \bar{p}(\cdot, 4T - t)$ for all $t \in [2T, 4T]$, we obtain that $\bar{p}$ is a solution of the wave equation in $[0, 4T]$. Finally, we extend the solution by periodicity with period $4T$. Noting that $\bar{p}(\cdot, 0) = \bar{p}(\cdot, 4T)$ and $\bar{p}_t(\cdot, 0) = \bar{p}_t(\cdot, 4T) = 0$, we obtain that $\bar{p}$ and $\bar{p}_t$ are continuous for all time and $\bar{p}$ satisfies the wave equation in $\mathbb{R}^3 \times \mathbb{R}_+$. This finishes our proof. □

Lemma 2. *There is a constant* C *and a pseudo-differential operator* $\mathbf{K}_T$ *of order at most* -1 *such that*

$$ \forall f \in L^2(B(0, R)) \colon \quad \|f\|_{L^2(\Omega)} \leq 2 \left(\|\mathbf{W}_T f\|_{L^2(\mathbb{R}^3)} + \|\mathbf{K}_T f\|_{L^2(B(0,R))} \right). $$

Proof. Let p denote the solution of wave Equations (1)–(3) and recall the parametrix formula $p(\mathbf{y}, t) = \frac{1}{(2\pi)^3} \sum_{\sigma = \pm} \int_{\mathbb{R}^3} a_\sigma(\mathbf{y}, t, \xi) e^{i\phi_\pm(\mathbf{y}, T, \xi)} \hat{f}(\xi) \, d\xi = \sum_{\sigma = \pm} p_\sigma(\mathbf{y}, t)$; see [47]. Here, the phase function $\phi_\pm$ solves the eikonal equation $\partial_t \phi_\pm(\mathbf{y}, t, \xi) \pm c(\mathbf{y}) |\nabla_\mathbf{y} \phi_\pm(\mathbf{y}, t, \xi)| = 0$ for $(\mathbf{y}, t) \in \mathbb{R}^3 \times \mathbb{R}_+$ with the initial condition $\phi_\pm(\mathbf{x}, 0, \xi) = \mathbf{x} \cdot \xi$. The amplitude function is a classical symbol $a_\pm(\mathbf{y}, t, \xi) = \sum_{k=0}^\infty a_{-k,\pm}(\mathbf{y}, t, \xi)$, where a_{-k} is homogeneous of order $-k$ in ξ. Its leading term $a_{0,\pm}$ satisfies the transport equation

$$\left(\partial_t \phi_\pm(\mathbf{y}, t, \xi) \partial_t - c^2(\mathbf{y}) \nabla_\mathbf{y} \phi_\pm(\mathbf{y}, t, \xi) \cdot \nabla_\mathbf{y} + C_\pm(\mathbf{y}, t, \xi) \right) a_{0,\pm}(\mathbf{y}, t, \xi) = 0, \tag{12}$$

with the initial condition $a_{\pm,0}(\mathbf{x}, 0, \xi) = 1/2$, see [41]. Here, $C(\mathbf{y}, \xi, t)$ only depends on the sound speed c and the phase function $\phi_\pm$. Let us denote by $\gamma_{\mathbf{x}, \xi}$ the unit speed geodesics originating at $\mathbf{x}$ along the direction ξ. Then, $\gamma_{\mathbf{x}, \xi}$ is a characteristics curve of the above transport equation; that is, (12) reduces to a homogeneous ODE on each geodesic curve.

We then write

$$\mathbf{W}_T(f)(\mathbf{y}) \quad = \quad \frac{1}{(2\pi)^3} \sum_{\sigma=\pm} \int_{\mathbb{R}^3} a_\sigma(\mathbf{y}, T, \xi) e^{i\phi_\pm(\mathbf{y}, T, \xi)} \hat{f}(\xi) \, d\xi = \sum_{\sigma=\pm} \mathbf{W}_\sigma(f)(\mathbf{y}).$$

Each operator $\mathbf{W}_\pm$ is a Fourier integral operator (FIO) with the canonical relation given by the pairs $(\mathbf{y}_\pm, \lambda\eta_\pm; \mathbf{x}, \lambda\xi)$ for any $\lambda \in \mathbb{R}$, ξ, η unit vectors, $\mathbf{y}_\pm = \gamma_{\mathbf{x}, \xi}(\pm T)$, and $\eta_\pm = \dot{\gamma}_{\mathbf{x}, \xi}(\pm T)$. Let $\mathbb{R}^3$ be equipped with the metrics $c^{-2}(\mathbf{x}) \, d\mathbf{x}^2$. Then, $(\mathbf{y}_\pm, \eta_\pm)$ is obtained by translating $(\mathbf{x}, \xi)$ on the geodesic $\gamma_{\mathbf{x}, \pm\xi}$ by the distance T. From the initial condition of $\phi_\pm$ and $a_{0,\pm}$ we see that, up to lower order terms,

$$p_-(\mathbf{x}, 0) = p_+(\mathbf{x}, 0) = \frac{1}{2} f(\mathbf{x}). \tag{13}$$

Heuristically, under Equations (1)–(3), each singularity of f at $(\mathbf{x}, \xi)$ is broken into two equal parts. They propagate along the geodesic $\gamma_{\mathbf{x}, \xi}$ in the opposite directions $\pm\xi$ to generate a singularity of $\mathbf{W}_T(f)$ at $(\mathbf{y}_\pm, \eta_\pm)$.

From the standard theory of FIOs (see [50]), the adjoint $\mathbf{W}_\pm^*$ translates $(\mathbf{y}_\pm, \eta_\pm)$ back to $(\mathbf{x}, \xi)$ and $\mathbf{W}_\pm^* \mathbf{W}_\pm$ is a pseudo differential operator. On the other hand, $\mathbf{W}_\mp^* \mathbf{W}_\pm$ is a FIO whose canonical relation consists of the pairs $(\mathbf{y}, \eta; \mathbf{x}, \xi)$ given by $\mathbf{y} = \gamma_{\mathbf{x}, \xi}(\pm 2T)$, and $\eta = \dot{\gamma}_{\mathbf{x}, \xi}(\pm 2T)$. That is, $\mathbf{W}_\pm^* \mathbf{W}_\mp$ is an infinitely smoothing operator on B. Therefore, microlocally, we can write $\mathbf{W}_T^* \mathbf{W}_T f = \mathbf{W}_+^* \mathbf{W}_+(f) + \mathbf{W}_-^* \mathbf{W}_-(f)$. We will show that the principal symbol $\theta_\pm(\mathbf{x}, \xi)$ of $\mathbf{W}_\pm^* \mathbf{W}_\pm$ satisfies $\theta_\pm(\mathbf{x}, \xi) = 1/4$. This result can be intuitively understood as follows. Let us consider $\mathbf{W}_+^* \mathbf{W}_+$ and a singularity of f at $(\mathbf{x}, \xi)$. Under Equations (1)–(3), half of this singularity propagates into the direction ξ (corresponding to the function p_+). At the moment $t = T$, it is transformed to a singularity of $\mathbf{W}_+(f) = p_+(T)$ at $(\mathbf{y}_+, \eta_+)$. Under the adjoint Equation (15), half of this singularity propagates back to $(\mathbf{x}, \xi)$ at $t = 0$ to generate a singularity of $\mathbf{W}_+^* \mathbf{W}_+(f)$. It is natural to believe that this recovered singularity is $1/4$ of the original singularity of f (due to twice splitting, as described). The proof below verifies this intuition.

Indeed, denote by q_+ the solution of the time-reversed wave equation, e.g., Equation (15), with the initial condition given by $g_+ = \mathbf{W}_+(f)$. Then, by definition (see Theorem 2) $\mathbf{W}_T^* g_+ = q_+(\cdot, 0)|_B$. The solution q_+ can be decomposed into the sum $q_+ = q_0 + q_1$. Here, q_0, q_1, up to smooth terms, are solutions of the wave equations in $\mathbb{R}^3 \times (0, T)$ and satisfy $q_0(\cdot, 0) = \mathbf{W}_+^*(g_+), q_1(\cdot, 0) = \mathbf{W}_-^*(g_+)$. We are only concerned with q_0 since it defines $\mathbf{W}_+^* \mathbf{W}_+ f = q_0(\cdot, 0)$. We can write

$$q_0(\mathbf{y}, t) = \frac{1}{(2\pi)^3} \int_{\mathbb{R}^3} b(\mathbf{y}, t, \xi) e^{i\phi_+(\mathbf{y}, t, \xi)} \hat{f}(\xi) \, d\xi. \tag{14}$$

Let b_0 be the principal part of b. Then, the principal symbol θ_+ of $\mathbf{W}_+^* \mathbf{W}_+$ is given by $\theta_+(\mathbf{x}, \xi) = b_0(\mathbf{x}, 0, \xi)$. We note that b_0 satisfies the same equation as $a_{0,+}$ (see (12)). Therefore, on each bicharacteristic curve the ratio $b_0/a_{0,+}$ is constant which implies $b_0(\mathbf{x}, 0, \xi) = $

$a_{+,0}(\mathbf{x}, 0, \xi)b_0(\mathbf{y}_+, T, \eta_+)/a_{+,0}(\mathbf{y}_+, T, \eta_+)$. Similar to the argument below Equation (13), up to lower order terms, we have

$$q_0(\mathbf{y}_+, T) = \frac{1}{2}g_+(\mathbf{y}_+, T) = \frac{1}{(2\pi)^3}\int_{\mathbb{R}^3}\frac{1}{2}a_+(\mathbf{y}_+, T, \xi)e^{i\phi_+(\mathbf{y}_+, T, \xi)}\hat{f}(\xi)\, d\xi.$$

This and Equation (14) implies that $b_0(\mathbf{y}_+, T, \xi) = a_{+,0}(\mathbf{y}_+, T, \xi)/2$. Therefore, we obtain $b_0(\mathbf{x}, 0, \xi) = a_{+,0}(\mathbf{x}, 0, \xi)/2 = 1/4$. Combining with a similar argument for $\mathbf{W}_-^*\mathbf{W}_-$, we obtain that the principal symbol of $\mathbf{W}_T^*\mathbf{W}_T$ is $\theta(\mathbf{x}, \xi) = 1/2$. That is, $\mathbf{W}_T^*\mathbf{W}_T = \mathbf{I}/2 + \mathbf{K}_T$, where $\mathbf{K}_T$ is a pseudodifferential operator of order at most -1 and $\mathbf{I}$ is the identity. We have $(\mathbf{W}_T f, \mathbf{W}_T f) = (\mathbf{W}_T^*\mathbf{W}_T f, f) = (f, f)/2 + (\mathbf{K}_T f, f)$ and therefore conclude $\|f\|_{L^2}^2 \leq 2(\|\mathbf{W}_T f\|_{L^2}^2 + \|\mathbf{K}_T f\|_{L^2}^2)$. $\square$

We are now ready to prove Theorem 1.

Proof of Theorem 1. Recall from Lemma 2 that $\|f\|_{L^2(B(0,R))} \leq 2(\|\mathbf{W}_T f\|_{L^2(\mathbb{R}^3)} + \|\mathbf{K}_T f\|_{L^2(B(0,R))})$ where $\mathbf{K}_T$ is a pseudo-differential operator of order at most -1. Since $\mathbf{K}_T$ is compact and $\mathbf{W}_T$ is injective, applying ([51] Theorem V.3.1), we obtain $\|f\|_{L^2(B(0,R))} \leq C\|\mathbf{W}_T f\|_{L^2(\mathbb{R}^3)}$ for some constant $C \in (0, \infty)$. This finishes our proof. $\square$

3.3. Continuous Adjoint Operator

Iterative methods for solving (9) require knowledge of the adjoint operator of $\mathbf{W}_T$. In this subsection, we compute the continuous adjoint operator $\mathbf{W}_T^*: L^2(\mathbb{R}^3) \to L^2(B(0, R))$ and prove that it is again given by the solution of a wave equation. More precisely, we have the following result.

Theorem 2. *Let $g \in C_0^\infty(\mathbb{R}^3)$, consider the time reversed final state problem for the wave equation,*

$$\begin{aligned} q_{tt}(\mathbf{x}, t) - c^2(\mathbf{x})\Delta q(\mathbf{x}, t) &= 0, & (\mathbf{x}, t) \in \mathbb{R}^3 \times (-\infty, T) \\ q(\mathbf{x}, T) &= g(\mathbf{x}), & \mathbf{x} \in \mathbb{R}^3 \\ q_t(\mathbf{x}, T) &= 0 & \mathbf{x} \in \mathbb{R}^3, \end{aligned} \tag{15}$$

and let $\chi_{B(0,R)}$ denote the indicator function of $B(0, R)$. Then, $\mathbf{W}_T^* g = \chi_{B(0,R)}\, q(\,\cdot\,, 0)$.

Proof. It is clearly sufficient to show $\mathbf{W}_T^* g = \chi_{B(0,R)}u_t(\,\cdot\,, 0)$, where u solves the wave equation $u_{tt} - c^2\Delta c = 0$ on $\mathbb{R}^3 \times (-\infty, T)$, with the final state conditions $(u(\,\cdot\,, T), u_t(\,\cdot\,, T)) = (0, f)$. Using the weak formulation (similar to [38]) for the wave equation, shows that $\int_0^T \int_{\mathbb{R}^3} c^{-2}(\mathbf{x})u_{tt}(\mathbf{x}, t)v(\mathbf{x}, t)\, dxdt + \int_0^T \int_{\mathbb{R}^3} \nabla u(\mathbf{x}, t) \cdot \nabla v(\mathbf{x}, t)\, dxdt = 0$ for $v \in C_0^\infty(\mathbb{R}^3)$. Two times integration by parts, rearranging terms and using the final state conditions for u yields $\int_{\mathbb{R}^3} c^{-2}(\mathbf{x})[f(\mathbf{x})v(\mathbf{x}, T) - u_t(\mathbf{x}, 0)v(\mathbf{x}, 0) + u(\mathbf{x}, 0)v_t(\mathbf{x}, 0)]\, dx = \int_0^T \int_{\mathbb{R}^3} u(\mathbf{x}, t)\left[c^{-2}(\mathbf{x})v_{tt}(\mathbf{x}, t) - \Delta v(\mathbf{x}, t)\right]dxdt$. By taking v as the solution of (1)–(3) this yields $\int_{\mathbb{R}^3} c^{-2}(\mathbf{x})g(\mathbf{x})\mathbf{W}_T(f)(\mathbf{x})\, dx = \int_{\mathbb{R}^3} c^{-2}(\mathbf{x})u_t(\mathbf{x}, 0)f(\mathbf{x})\, dx$, which implies $\mathbf{W}_T^* g = \chi_{B(0,R)}u_t(\,\cdot\,, 0) = \chi_{B(0,R)}q(\,\cdot\,, 0)$ and completes the proof. $\square$

We can reformulate the adjoint operator as follows.

Corollary 1. *For $g \in C_0^\infty(\mathbb{R}^3)$, let q be the solution of*

$$q_{tt}(\mathbf{x}, t) - c^2(\mathbf{x})\Delta q(\mathbf{x}, t) = 0 \quad \text{for } (\mathbf{x}, t) \in \mathbb{R}^3 \times (0, \infty) \tag{16}$$

with initial conditions $(q(\,\cdot\,, 0), q_t(\,\cdot\,, 0)) = (g, 0)$. Then, $\mathbf{W}_T^* g = \chi_{B(0,R)}\, q(\,\cdot\,, T)$.

Proof. Clearly q solves (16) with initial data $(q(\,\cdot\,, 0), q_t(\,\cdot\,, 0)) = (g, 0)$ if and only if $(x, t) \mapsto q(x, T - t)$ solves (15). Therefore, the claim follows from Theorem 2. $\square$

3.4. Application of the Steepest 6hhod

We propose solving the finite time wave inversion problem (9) via gradient type methods applied to residual functional $\Phi(f) := \frac{1}{2} \|\mathbf{W}_T f - h\|^2$. Using standard PAT data, various iterative methods for PAT accounting for variable sound speed have been investigated in [36–40]. In particular, in [40] the steepest descent method has been demonstrated to be numerically efficient and robust. For the results shown in this paper we will therefore use the steepest-descent method. Our numerical experiments confirm that the steepest-descent method is also efficient for FFD-PAT and the final time wave inversion problem, where it reads as follows.

Because of the injectivity and boundedness of $\mathbf{W}_T$, the sequence of iterates generated by Algorithm 1 converges to the unique solution of (9). The stability result (11) even implies that the steepest descent method converges linearly for FFD-PAT. More precisely, the sequence $(f_k)_{k \in \mathbb{N}}$ generated by Algorithm 1 satisfies the estimate $\|f_{k+1} - f\|_2 \le c^k \|f_0 - f\|_2^2$ for some constant $c \in (0, 1)$.

Algorithm 1: Steepest descent method for solving $\mathbf{W}_T f = g$.

(S1) Choose $a_\star \in (0, 1]$ and initialize $f_0 = 0$, $k \leftarrow 0$
(S2) While stopping criteria not satisfied, do

- $s_k = \mathbf{W}_T^*(\mathbf{W}_T f_k - g)$
- $a_k = a_\star \|s_k\|^2 / \|\mathbf{W}_T s_k\|^2$
- $f_{k+1} = f_k - a_k s_k$
- $k \leftarrow k + 1$

---

4. Numerical Simulations

In this section, we numerical present results using Algorithm 1 for FFD-PAT with variable sound speed. Recall that for constant speed of sound, reconstructions from experimental FFD-PAT data are shown in [30,31]. In the present study, we give numerical results for spatially varying speed of sound. Performing experimental measurements for samples with variable speed of sound and applying our algorithms to such data is subject of future research.

4.1. Discretization

To implement the steepest descent method (or other gradient type schemes), one has to discretize the final time wave operator $\mathbf{W}_T$ and its adjoint $\mathbf{W}_T^*$. For that purpose we solve the forward and adjoint wave Equations (9) and (16) on a cubical grid with side length $2L$ and nodes $\mathbf{x}[\mathbf{i}] := -(L, L, L) + \mathbf{i} 2L/N_x$ for $\mathbf{i} = (i_1, i_2, i_3) \in \{0, \dots, N_x - 1\}$ with the k-space method [52,53], which we briefly recall in Appendix A. Note that the implementation of the k-space method yields a $2L$-periodic solution. The parameter L is chosen such that $a + T < L$ which implies that inside $B(0, R)$, for $t \in [0, T]$, the solution of (16) coincides with its $2L$-periodic extension.

We denote by $\mathcal{F} \subseteq \mathbb{R}^{N_x \times N_x \times N_x}$ the set of all $\mathbf{f} \in \mathbb{R}^{N_x \times N_x \times N_x}$ with $\mathbf{f}[\mathbf{i}] = 0$ for $\mathbf{x}[\mathbf{i}] \notin B(0, R)$. The discretized versions of $\mathbf{W}_T$ and its adjoint $\mathbf{W}_T^*$ are defined by $\mathcal{W} : \mathcal{F} \to \mathbb{R}^{N_x \times N_x \times N_x} : \mathbf{f} \mapsto (\mathcal{K}\mathbf{f})(\cdot, N_t)$ and $\mathcal{W}^\mathsf{T} : \mathbb{R}^{N_x \times N_x \times N_x} \to \mathcal{F} : \mathbf{h} \mapsto \chi_R(\mathcal{K}\mathbf{h})(\cdot, N_t)$, respectively, where $\mathcal{K} : \mathbb{R}^{N \times N \times N} \to \mathbb{R}^{N_x \times N_x \times N_x \times (N+1)}$ denotes the discretized wave propagation defined by the k-space method using the discrete time steps jT/N_t for $0 \le j \le N_t$, and χ_R is the discretized indicator function of the ball $B(0, R)$. The linear projection $\mathbf{X}$ and its left inverse $\mathbf{X}^\sharp$ are implemented using the MATLAB build in functions for the Radon and inverse Radon transforms, respectively, with N_θ equally spaced projection angles covering $[0, \pi]$. The resulting discretized transforms are denoted by $\mathcal{X}$ and $\mathcal{X}^\sharp$, respectively.

4.2. Data Simulation

For the presented numerical results, we use $R = 0.4$, $T = 1$, $L = 1.5$, $N = 300$, $M = 300$ and collect FFD-data for $N_\theta = 300$ projection directions. As initial pressure $\mathbf{f}_\star$ we take the sum of three solid spheres, as show in the left picture in Figure 3. The used sound speed $\mathbf{c}_\star$ is shown in the right picture in Figure 3, and consist of two Gaussian peaks with opposite signs, added to the constant sound speed $c_0[i] = 1$. The top row in Figure 4 shows the simulated FFD-PAT data $\mathbf{g}_\star = \mathcal{X}\mathcal{W}\mathbf{f}_\star$ (left) and the noisy FFD-PAT data $\mathbf{g} := \mathbf{g}_\star + \text{noise}$ (right) at projection angle $\theta = 0$ for both cases. To generate the noisy FFD data we have added Gaussian white noise with a standard deviation of 10% of the maximal value of $\mathbf{g}_\star$, resulting in a relative L^2-data error $\|\text{noise}\| / \|\mathbf{g}_\star\| \simeq 85.3\%$. For the first reconstruction step we apply the inverse Radon transform to the FFD-PAT data, resulting in the approximate 3D final time pressure $\mathcal{X}^\sharp \mathbf{g} \simeq \mathcal{W}\mathbf{f}_\star$. The reconstruction of final time pressure from noisy data is shown in the bottom right image in Figure 4. For comparison purpose, the simulated final time pressure $\mathbf{h}_\star := \mathcal{W}\mathbf{f}_\star$ is shown in the bottom left image in Figure 4.

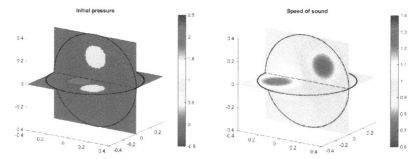

Figure 3. PHANTOM AND SOUND SPEED USED FOR THE NUMERICAL SIMULATIONS. (**Left**): Initial pressure $\mathbf{f}_\star$ (the phantom to be recovered). (**Right**): Spatially varying sound speed $c_\star$.

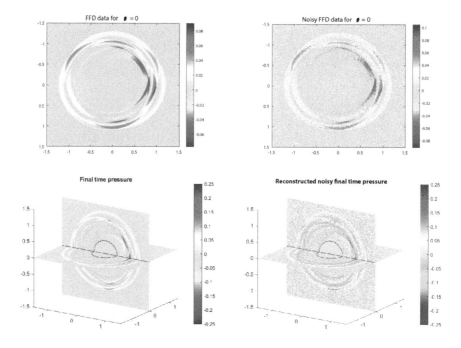

Figure 4. FFD-PAT MEASUREMENT DATA AND FINAL TIME PRESSURE. (**Top left**): Simulated FFD-PAT data $\mathbf{g}_\star$ for projection direction $\theta = 0$. (**Top right**): Noisy FFD-PAT data $\mathbf{g}$ for projection direction $\theta = 0$. (**Bottom left**): Simulated final time pressure $\mathbf{h}_\star = \mathcal{W}\mathbf{f}_\star$. (**Bottom right**): Recovered final time pressure $\mathcal{X}^\sharp \mathbf{g} \simeq \mathbf{h}_\star$ from noisy FFD-PAT data.

4.3. Reconstruction Results

To recover the initial pressure $\mathbf{f}_\star$, we apply the steepest descent method (Algorithm 1) with $a_\star = 0.8$ to the point-wise approximation $\mathbf{h} \simeq \mathcal{W}\mathbf{f}_\star$ that is recovered in step one. Reconstruction results after 5 iterations are shown in Figure 5. The top row shows reconstruction results from simulated data and the middle row shows reconstruction results from noisy data, both using the correct sound speed $\mathbf{c}_\star$ for the reconstruction process. One observes, that the reconstruction results from simulated as well as from noisy data are quite accurate. This demonstrates that the proposed algorithm is efficient, accurate and stable with respect to noise. In both cases, the whole reconstruction procedure takes about 90 min on a standard desktop PC.

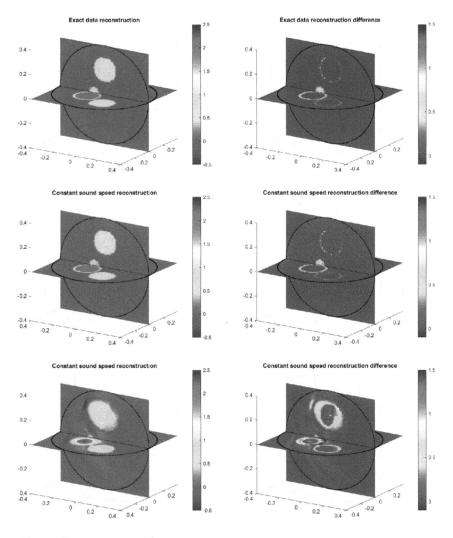

Figure 5. RECONSTRUCTIONS $\mathbf{f}_{\text{rec}}$ USING FIVE ITERATIONS OF THE STEEPEST DESCENT ALGORITHM (**LEFT**) AND THE DIFFERENCES $|\mathbf{f}_{\text{rec}} - \mathbf{f}_\star|$ TO THE TRUE PHANTOM (**RIGHT**). (**Top**): Using simulated data $\mathbf{g}_\star$ and correct sound speed $\mathbf{c}_\star$. (**Middle**): Using noisy data $\mathbf{g}$ and correct sound speed $\mathbf{c}_\star$. (**Bottom**): Using simulated data $\mathbf{g}_\star$ data and wrong sound speed $\mathbf{c}_0$.

The bottom row in Figure 5 shows reconstruction results using the wrong constant sound speed c_0 in Algorithm 1 (while data generation still uses the inhomogeneous sound speed $c_\star$). In this case, the reconstruction error is much larger, showing the relevance of integrating sound speed variations in image reconstruction. Finally, in Figure 6 we show reconstruction results with the previous Fourier method that has been derived in [31] under a constant sound speed assumption. The results with the Fourier method show a large reconstruction error and are very similar to the reconstruction result using the steepest descent method assuming constant sound speed. This demonstrates that the artifacts in both cases are due to the wrong wave propagation model, which further supports the importance of taking sound speed variations into account in FFD-PAT image reconstruction.

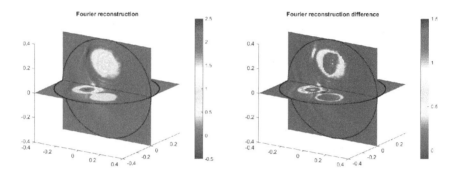

Figure 6. RECONSTRUCTION $\mathbf{f}_{\mathrm{rec}}$ (LEFT) AND THE DIFFERENCE $|\mathbf{f}_{\mathrm{rec}} - \mathbf{f}_\star|$ (RIGHT) USING THE FOURIER ALGORITHM. Note that the Fourier algorithm assumes constant sound speed c_0 in the image reconstruction, similar to the results shown in the bottom row of Figure 5.

4.4. Quantitative Error Analysis

For all results shown in Figure 5, we stopped the steepest decent method after five iterations, since the reconstructions did not significantly improve by using more iterations. In fact, even after a single iteration, the ℓ^2-reconstruction error and the ℓ^2-residual error are quite close to their minimal values. This phenomenon is investigated in a more quantitative manner in Figure 7. The images on the left-hand side show the evaluation of the relative L^2-reconstruction error $\|\mathbf{f}_k - \mathbf{f}_\star\| / \|\mathbf{f}_\star\|$ in dependence of the iteration index k. The images on the right show the relative residual errors $\|\mathcal{W}\mathbf{f}_k - \mathbf{h}_\star\| / \|\mathbf{h}_\star\|$. The rapid convergence of both error metrics indicates that the finite time inversion problem is well conditioned, which is also suggested by our theoretical analysis presented in Section 3. The relative data errors, residuals and reconstruction errors for all reconstructions shown above are summarized in Table 1.

Table 1. RELATIVE ERROR METRICS FOR THE FFD DATA AND RECONSTRUCTIONS. The first column shows the relative ℓ^2-norm of the noise added to the data. The second column shows the relative ℓ^2-reconstruction error after step one (not applicable to Fourier reconstruction). The third column shows the relative residual error which are minimized in step 2 of the iterative algorithms. The last column shows the relative ℓ^2-error of final reconstruction.

	$\frac{\|\mathbf{g}-\mathbf{g}_\star\|}{\|\mathbf{g}_\star\|}$ Noise	$\frac{\|\mathcal{X}^\sharp \mathbf{g}-\mathbf{h}_\star\|}{\|\mathbf{h}_\star\|}$ Error (after Step 1)	$\frac{\|\mathcal{W}\mathbf{f}_{\mathrm{rec}}-\mathbf{h}_\star\|}{\|\mathbf{h}_\star\|}$ Residual (for Step 2)	$\frac{\|\mathbf{f}_{\mathrm{rec}}-\mathbf{f}_\star\|}{\|\mathbf{f}_\star\|}$ Error (after Step 2)
Correct SS (simulated data)	0	0.208	0.052	0.170
Correct SS (noisy data)	0.853	1.760	0.874	0.227
Wrong SS (simulated data)	0	0.208	0.510	0.560
Fourier method [31]	0	-	-	0.564

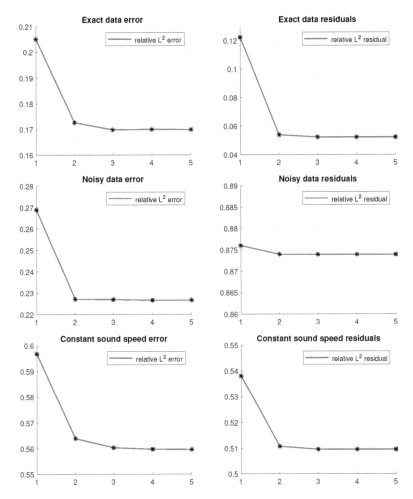

Figure 7. RELATIVE RECONSTRUCTION ERROR (**LEFT**) AND RELATIVE RESIDUALS (**RIGHT**). (**Top**): Simulated data and correct sound speed. (**Middle**): Noisy data and correct sound speed. (**Bottom**): Simulated data and wrong sound speed (constant and equal to one).

5. Conclusions

In this paper, we investigated FFD-PAT, where projection data of acoustic pressure are measured. For the first time, the variable speed of sound has been taken into account for this kind of setup. We developed a two-step reconstruction procedure that recovers the 3D pressure $p(\,\cdot\,,T)$ in a first step, which is then used as input for a finite time wave inversion problem in a second step. As the main theoretical contribution of this paper we prove uniqueness and stability results for the finite time wave inversion problem. For the actual solution, we propose the steepest descent iteration which we found to be numerically efficient and stable for FFD-PAT. Moreover, our numerical results demonstrate that ignoring sound speed variations significantly degrades the reconstruction quality. We point out that the novelties of the paper are the development of a reconstruction method together with a mathematical uniqueness and stability analysis of the arising final time wave inversion problem for FFD-PAT with variable sound speed. The experimental feasibility of FFD-PAT has been demonstrated previously in [30,31]. In future work we will perform experiments using FFD-PAT with spatially variable sound speed and experimentally verify our two-step algorithm. Moreover, in upcoming work we will study limited view problems for FFD-PAT which naturally arise in applications, for instance in the case of breast imaging.

Author Contributions: M.H. and G.Z. developed the reconstruction algorithms, the numerical implementation and performed the numerical experiments; L.N. established the uniqueness and stability results; M.H., G.Z., L.N. and R.N. wrote the paper; M.H. and R.N. supervised the project.

Funding: G.Z. and M.H. acknowledge support of the Austrian Science Fund (FWF), project P 30747-N32. The research of L.N. is supported by the National Science Foundation (NSF) Grants DMS 1212125 and DMS 1616904. The work of R.N. has been supported by the FWF, project P 28032.

Conflicts of Interest: The authors declare no conflicts of interest.

Appendix A. k-Space Method

We briefly describe the k-space method for the 3D wave Equations (1)–(3) as we use it for the numerical computation of $\mathbf{W}_T$ and $\mathbf{W}_T^*$. The k-space method is an attractive alternative to standard methods using finite differences, finite elements or pseudospectral methods, since it does not suffer from numerical dispersion [52,53]. It utilizes the decomposition $p(\mathbf{x}, t) = w(\mathbf{x}, t) - v(\mathbf{x}, t)$, where v, w are defined by $w(\mathbf{x}, t) := (c_0^2/c^2(\mathbf{x})) p(\mathbf{x}, t)$ and $v(\mathbf{x}, t) = (c_0^2/c^2(\mathbf{x}) - 1)p(\mathbf{x}, t)$ where $c_0 := \max\{c(\mathbf{x}) \mid \mathbf{x} \in \mathbb{R}^3\}$ denotes maximal speed of sound. It can be checked that with this definition of v and w the wave equation with variable speed of sound splits into the system $w_{tt}(\mathbf{x}, t) - c_0^2 \Delta w(\mathbf{x}, t) = -c_0^2 \Delta v(\mathbf{x}, t)$ and $v(\mathbf{x}, t) = \frac{c_0^2 - c^2(\mathbf{x})}{c_0^2} w(\mathbf{x}, t)$. In the k-space method we use the time stepping formula

$$w(\mathbf{x}, t + h_t) = 2w(\mathbf{x}, t) - w(\mathbf{x}, t - h_t) - 4\mathbf{F}_\xi^{-1}\left\{\sin\left(\frac{c_0|\xi|h_t}{2}\right)^2 \mathbf{F}_\mathbf{x}\{w(\mathbf{x}, t) - v(\mathbf{x}, t)\}\right\}, \qquad \text{(A1)}$$

where $\mathbf{F}_\mathbf{x}$ and $\mathbf{F}_\xi^{-1}$ denote the Fourier transform and its inverse with respect to space and frequency variables $\mathbf{x}$ and ξ and h_t is the time step size. This equivalent formulation motivates the following algorithm for numerically solving the wave equation.

Algorithm A1: k-space method for numerically solving (1)–(3).

(S1) Define initial conditions:

$$w(\mathbf{x}, -h_t) = w(\mathbf{x}, 0) = c_0^2/c^2(\mathbf{x})f(\mathbf{x}),$$
$$v(\mathbf{x}, 0) = (c_0^2/c^2(\mathbf{x}) - 1)f(\mathbf{x})$$

(S2) Set $t = 0$
(S3) Compute $w(\mathbf{x}, t + h_t)$ according to Equation (A1)
(S4) Compute $v(\mathbf{x}, t + h_t) := (c^2(\mathbf{x})/c_0^2 - 1)w(\mathbf{x}, h_t)$
(S5) Compute $p(\mathbf{x}, t + h_t) := w(\mathbf{x}, t + h_t) - w(\mathbf{x}, t + h_t)$
(S6) Substitute t by $t + h_t$ and go back to step (3).

References

1. Beard, P. Biomedical photoacoustic imaging. *Interface Focus* **2011**, *1*, 602–631. [CrossRef]
2. Kruger, R.A.; Kopecky, K.K.; Aisen, A.M.; Reinecke, R.D.; Kruger, G.A.; Kiser, W.L. Thermoacoustic CT with Radio waves: A medical imaging paradigm. *Radiology* **1999**, *200*, 275–278. [CrossRef] [PubMed]
3. Wang, L.V. Multiscale photoacoustic microscopy and computed tomography. *Nat. Photonics* **2009**, *3*, 503–509. [CrossRef]
4. Wang, K.; Anastasio, M. Photoacoustic and thermoacoustic tomography: Image formation principles. In *Handbook of Mathematical Methods in Imaging*; Springer: Berlin, Germany, 2011; pp. 781–815.
5. Xu, M.; Wang, L.V. Photoacoustic imaging in biomedicine. *Rev. Sci. Instrum.* **2006**, *77*, 041101. [CrossRef]

6. Agranovsky, M.; Kuchment, P.; Kunyansky, L. On reconstruction formulas and algorithms for the thermoacoustic tomography. In *Photoacoustic Imaging and Spectroscopy*; Wang, L.V., Ed.; CRC Press: Boca Raton, FL, USA, 2009; pp. 89–101.
7. Filbir, F.; Kunis, S.; Seyfried, R. Effective discretization of direct reconstruction schemes for photoacoustic imaging in spherical geometries. *SIAM J. Numer. Anal.* **2014**, *52*, 2722–2742. [CrossRef]
8. Finch, D. The spherical mean value operator with centers on a sphere. *Inverse Probl.* **2007**, *23*, 37–49. [CrossRef]
9. Finch, D.; Haltmeier, M.; Rakesh Inversion of spherical means and the wave equation in even dimensions. *SIAM J. Appl. Math.* **2007**, *68*, 392–412. [CrossRef]
10. Finch, D.; Patch, S.K. Determining a function from its mean values over a family of spheres. *SIAM J. Math. Anal.* **2004**, *35*, 1213–1240. [CrossRef]
11. Haltmeier, M. Universal inversion formulas for recovering a function from spherical means. *SIAM J. Math. Anal.* **2014**, *46*, 214–232. [CrossRef]
12. Haltmeier, M. Exact Reconstruction Formula for the Spherical Mean Radon Transform on Ellipsoids. *Inverse Probl.* **2014**, *30*, 035001. [CrossRef]
13. Haltmeier, M.; Pereverzyev, S., Jr. The universal back-projection formula for spherical means and the wave equation on certain quadric hypersurfaces. *J. Math. Anal. Appl.* **2015**, *429*, 366–382. [CrossRef]
14. Haltmeier, M.; Schuster, T.; Scherzer, O. Filtered backprojection for thermoacoustic computed tomography in spherical geometry. *Math. Methods Appl. Sci.* **2005**, *28*, 1919–1937. [CrossRef]
15. Kuchment, P.; Kunyansky, L.A. Mathematics of thermoacoustic and photoacoustic tomography. *Eur. J. Appl. Math.* **2008**, *19*, 191–224. [CrossRef]
16. Kunyansky, L.A. Explicit inversion formulae for the spherical mean Radon transform. *Inverse Probl.* **2007**, *23*, 373–383. [CrossRef]
17. Kunyansky, L.A. A series solution and a fast algorithm for the inversion of the spherical mean Radon transform. *Inverse Probl.* **2007**, *23*, S11–S20. [CrossRef]
18. Natterer, F. Photo-acoustic inversion in convex domains. *Inverse Probl. Imaging* **2012**, *6*, 315–320. [CrossRef]
19. Nguyen, L.V. A family of inversion formulas for thermoacoustic tomography. *Inverse Probl. Imaging* **2009**, *3*, 649–675. [CrossRef]
20. Xu, M.; Wang, L.V. Universal back-projection algorithm for photoacoustic computed tomography. *Phys. Rev. E* **2005**, *71*, 016706. [CrossRef] [PubMed]
21. Haltmeier, M.; Zangerl, G. Spatial resolution in photoacoustic tomography: Effects of detector size and detector bandwidth. *Inverse Probl.* **2010**, *26*, 125002. [CrossRef]
22. Roitner, H.; Haltmeier, M.; Nuster, R.; O'Leary, D.P.; Berer, T.; Paltauf, G.; Grün, H.; Burgholzer, P. Deblurring algorithms accounting for the finite detector size in photoacoustic tomography. *J. Biomed. Opt.* **2014**, *19*, 056011. [CrossRef] [PubMed]
23. Rosenthal, A.; Ntziachristos, V.; Razansky, D. Model-based optoacoustic inversion with arbitrary-shape detectors. *Med. Phys.* **2011**, *38*, 4285–4295. [CrossRef]
24. Wang, L.V. An imaging model incorporating ultrasonic transducer properties for three-dimensional optoacoustic tomography. *IEEE Trans. Med. Imaging* **2011**, *30*, 203–214. [CrossRef] [PubMed]
25. Burgholzer, P.; Hofer, C.; Paltauf, G.; Haltmeier, M.; Scherzer, O. Thermoacoustic tomography with integrating area and line detectors. *IEEE Trans. Ultrason. Ferroelectr. Freq. Control* **2005**, *52*, 1577–1583. [CrossRef] [PubMed]
26. Burgholzer, P.; Bauer-Marschallinger, J.; Grün, H.; Haltmeier, M.; Paltauf, G. Temporal back-projection algorithms for photoacoustic tomography with integrating line detectors. *Inverse Probl.* **2007**, *23*, S65–S80. [CrossRef]
27. Haltmeier, M.; Scherzer, O.; Burgholzer, P.; Paltauf, G. Thermoacoustic computed tomography with large planar receivers. *Inverse Probl.* **2004**, *20*, 1663. [CrossRef]
28. Paltauf, G.; Nuster, R.; Haltmeier, M. Experimental evaluation of reconstruction algorithms for limited view photoacoustic tomography with line detectors. *Inverse Probl.* **2007**, *23*, S81–S94. [CrossRef]
29. Zangerl, G.; Scherzer, O.; Haltmeier, M. Exact series reconstruction in photoacoustic tomography with circular integrating detectors. *Commun. Math. Sci.* **2009**, *7*, 665–678.
30. Nuster, R.; Zangerl, G.; Haltmeier, M.; Paltauf, G. Full field detection in photoacoustic tomography. *Opt. Express* **2010**, *18*, 6288–6299. [CrossRef]

31. Nuster, R.; Slezak, P.; Paltauf, G. High resolution three-dimensional photoacoustic tomography with CCD-camera based ultrasound detection. *Biomed. Opt. Express* **2014**, *5*, 2635–2647. [CrossRef] [PubMed]
32. Niederhauser, J.J.; Weber, D.F.H.P.; Frenz, M. Real-time optoacoustic imaging using a Schlieren transducer. *Appl. Phys. Lett.* **2002**, *81*, 571–573. [CrossRef]
33. Niederhauser, J.J.; Jäger, M.; Frenz, M. Real-time three-dimensional optoacoustic imaging using an acoustic lens system. *Appl. Phys. Lett.* **2004**, *85*, 846–848. [CrossRef]
34. Jin, X.; Wang, L.V. Thermoacoustic tomography with correction for acoustic speed variations. *Phys. Med. Biol.* **2006**, *51*, 6437. [CrossRef] [PubMed]
35. Ku, G.; Fornage, B.D.; Xing, J.; Xu, M.; Hunt, K.K.; Wang, L.V. Thermoacoustic and photoacoustic tomography of thick biological tissues toward breast imaging. *Med. Phys.* **1995**, *22*, 1605–1609. [CrossRef] [PubMed]
36. Belhachmi, Z.; Glatz, T.; Scherzer, O. A direct method for photoacoustic tomography with inhomogeneous sound speed. *Inverse Probl.* **2016**, *32*, 045005. [CrossRef]
37. Arridge, S.R.; Betcke, M.M.; Cox, B.T.; Lucka, F.; Treeby, B.E. On the adjoint operator in photoacoustic tomography. *Inverse Probl.* **2016**, *32*, 115012. [CrossRef]
38. Haltmeier, M.; Nguyen, L.V. Analysis of Iterative Methods in Photoacoustic Tomography with variable Sound Speed. *SIAM J. Imaging Sci.* **2017**, *19*, 751–781. [CrossRef]
39. Huang, C.; Wang, K.; Nie, L.; Wang, L.V.; Anastasio, M.A. Full-wave iterative image reconstruction in photoacoustic tomography with acoustically inhomogeneous media. *IEEE Trans. Med. Imaging* **2013**, *32*, 1097–1110. [CrossRef] [PubMed]
40. Nguyen, L.V.; Haltmeier, M. Reconstruction algorithms for photoacoustic tomography in heterogenous damping media. *arXiv* **2018**, arXiv:1808.06176v1.
41. Stefanov, P.; Uhlmann, G. Thermoacoustic tomography with variable sound speed. *Inverse Probl.* **2009**, *25*, 075011. [CrossRef]
42. Natterer, F. *The Mathematics of Computerized Tomography*; SIAM: Philadelphia, PA, USA, 1986.
43. Kuchment, P. *The Radon Transform and Medical Imaging*; SIAM: Philadelphia, PA, USA, 2014; Volume 85.
44. Quinto, E. Singular value decompositions and inversion methods for the exterior Radon transform and a spherical transform. *J. Math. Anal. Appl.* **1983**, *95*, 437–448. [CrossRef]
45. Quinto, E. Tomographic reconstructions from incomplete data-numerical inversion of the exterior Radon transform. *Inverse Probl.* **1988**, *4*, 867. [CrossRef]
46. Vainberg, B. On the short wave asymptotic behaviour of solutions of stationary problems and the asymptotic behaviour as $t \to \infty$ of solutions of non-stationary problems. *Rus. Math. Surv.* **1975**, *30*, 1–58. [CrossRef]
47. Tréves, F. *Introduction to Pseudodifferential and Fourier Integral Operators Volume 2: Fourier Integral Operators*; Springer Science & Business Media: Berlin, Germany, 1980.
48. John, F. Partial Differential Equations. In *Applied Mathematical Sciences*, 4th ed.; Springer: New York, NY, USA, 1982; Volume 1.
49. Nguyen, L.V. On singularities and instability of reconstruction in thermoacoustic tomography. *Tomogr. Inverse Transp. Theory Contemp. Math.* **2011**, *559*, 163–170.
50. Hörmander, L. Fourier integral operators. I. *Acta Math.* **1971**, *127*, 79–183. [CrossRef]
51. Taylor, M.E. *Pseudodifferential Operators, Volume 34 of Princeton Mathematical Series*; Princeton University Press: Princeton, NJ, USA; Springer Science & Business Media: Berlin, Germany, 1981
52. Cox, B.T.; Kara, S.; Arridge, S.R.; Beard, P.C. k-space propagation models for acoustically heterogeneous media: Application to biomedical photoacoustics. *J. Acoust. Soc. Am.* **2007**, *121*, 3453–3464. [CrossRef]
53. Mast, T.D.; Souriau, L.P.; Liu, D-.D.; Tabei, M.; Nachman, A.I.; Waag, R.C. A k-space method for large-scale models of wave propagation in tissue. *IEEE Trans. Ultrason. Ferroelectr. Freq. Control* **2002**, *48*, 341–354. [CrossRef]

Article

Development of Low-Cost Fast Photoacoustic Computed Tomography: System Characterization and Phantom Study

Mohsin Zafar [1], Karl Kratkiewicz [1], Rayyan Manwar [1] and Mohammad Avanaki [1,2,3,*]

1   Department of Biomedical Engineering, Wayne State University, Detroit, MI 48201, USA;
    mohsin.zafar@wayne.edu (M.Z.); karl.kratkiewicz@wayne.edu (K.K.); r.manwar@wayne.edu (R.M.)
2   Department of Neurology, Wayne State University School of Medicine, Detroit, MI 48201, USA
3   Barbara Ann Karmanos Cancer Institute, Detroit, MI 48201, USA
*   Correspondence: mrn.avanaki@wayne.edu; Tel.: +1-313-577-0703

Received: 4 December 2018; Accepted: 12 January 2019; Published: 22 January 2019

Abstract: A low-cost Photoacoustic Computed Tomography (PACT) system consisting of 16 single-element transducers has been developed. Our design proposes a fast rotating mechanism of 360° rotation around the imaging target, generating comparable images to those produced by large-number-element (e.g., 512, 1024, etc.) ring-array PACT systems. The 2D images with a temporal resolution of 1.5 s and a spatial resolution of 240 μm were achieved. The performance of the proposed system was evaluated by imaging complex phantom. The purpose of the proposed development is to provide researchers a low-cost alternative 2D photoacoustic computed tomography system with comparable resolution to the current high performance expensive ring-array PACT systems.

Keywords: Photoacoustic Computed Tomography (PACT); ring array; fast imaging; low cost

1. Introduction

Photoacoustic imaging (PAI) is a promising imaging technique that combines higher imaging contrast of optical imaging with deeper penetration of ultrasonic imaging [1–3]. A nanosecond pulsed laser deposits energy onto a light absorbing sample, causing a local temperature increase and subsequent thermal expansion through the thermoacoustic effect [2,4–8]. This expansion causes a localized pressure increase, which propagates from the sample to be imaged, by an ultrasound transducer, similar to a traditional ultrasound detection system. PAI has gained popularity compared to other optical imaging techniques mainly due to its simultaneous great penetration depth and adequate resolution. PAI utilizes the deposition of diffusive photons and therefore can image at much greater depths [9,10] as compared to other high-resolution optical imaging modalities such as optical coherence tomography (OCT) [11,12]; OCT's portability and high resolution however stand out in relevant applications [13,14]. As opposed to positron emission tomography (PET) and computed tomography (CT) that require the use of ionizing radiation, PAI uses safe, non-ionizing, visible and near-infrared light. Compared to magnetic resonance imaging (MRI) that requires sophisticated, large, and expensive designs, PAI is portable and cost-effective. PAI is capable of both functional and molecular imaging. PAI has proved to have a superior specificity in clinical diagnosis over conventional ultrasound imaging [15–18]. The wavelength dependence nature of PAI allows imaging of various molecules in biological tissues using endogenous and exogenous contrast agents [19]; e.g., tissue oxygen saturation can be measured using multispectral PAI [20–22]. The molecular imaging capability of PAI enables anatomical, and functional imaging of organelles, organs or even whole body of small animals, as well as many applications in human tissue analysis [19,23,24]. These advanced features have allowed PAI to play an important role in small animal imaging though different human diseases can be comprehensively studied. Furthermore, PAI can be integrated with other imaging modalities to generate multi-modality images that provide complementary information of the tissue.

There are two major implementations for PA imaging, i.e., photoacoustic computed tomography (PACT) and photoacoustic microscopy (PAM). PACT is used for deep tissue imaging applications where coarse resolution is acceptable. In PACT, high energy pulsed laser light is diffused to illuminate the entire tissue, which then generates photoacoustic waves. The waves around the tissue are collected by wideband ultrasound transducers, placed in the same plane, at radially symmetric angles, and ideally equidistant from the imaging target [25]. The detection scheme can be realized either by a single ultrasound transducer rotating around the sample, or a stationary ring array of 128, 256, or greater number of transducer elements [7,26,27]. Such configuration offers a fast image acquisition, but requires an expensive hardware and data acquisition unit. Moreover, PACT systems typically utilize bulky nanosecond pulsed laser sources, sometimes with an optical parametric oscillator (OPO). In addition to the light source, a major cost in these systems comes from the transducer arrays.

Some of the PACT systems developed are as follows. A 128-channel curved array PACT system was reported in [28] with 15–30 s rotation time to acquire a 2D image of the sample with a resolution of about 200 μm; rotating the animal for small animal in-vivo imaging is not recommended, also the long acquisition time does not allow imaging the hemodynamic changes (such changes occur with the frame rate of 2.5 Hz in small animals [29]). Gamelin et al. developed a fast, expensive, sophisticated 512-elements full ring PACT system that acquired images in less than one second [30]. The complexity and high cost of such a system prevents most researchers from developing PACT systems of their own. Therefore, strategies of decreasing the cost of PACT are needed. Some researchers have attempted to reduce the cost of ultrasound detection units by using partial view detection with reflectors [31] or using sparse arrays along with a compressed sensing reconstruction algorithm [32].

In [33], a multispectral PACT system was implemented where 8 single element transducers were utilized in an arc fashion. The system took around 30 s to acquire an image. In [25], a PACT system was developed using only two unfocused transducers, and a complete rotation was performed in 3s. In [34], a single element focused transducer-based PACT system was developed for in-vivo rat brain imaging. The total acquisition time was 16 minutes with a scanning step size of 1.5°. In [35], a spherical PACT system was developed to image zebra fish. The scanning step size was 0.75° with 480 scanning positions and 24 s acquisition time for each frame.

In this article, we present the development and characterization of a low-cost, fast PACT system that has been optimized to achieve 1.5 s temporal resolution with a 240 μm spatial resolution. The low number of elements (16) reduces the cost of the ultrasound unit as well as the need for an expensive DAQ. To reduce the cost of illumination, a Q-switched Nd:YAG laser was used. This comparatively lower-cost system provides fast, high-resolution images to potentially be used for small animal imaging. The performance of the system has been evaluated by imaging complex phantoms.

2. Materials and Methods

Figure 1 illustrates an experimental set up and architecture of our low-cost photoacoustic tomography system. A Q-switched Nd:YAG laser (NL231-50-SH, EKSPLA, Vilnius, Lithuania) with a pulse width of ~5 ns and pulse repetition rate of 50 Hz was used. The laser produces 532 nm and 1064 nm wavelengths.

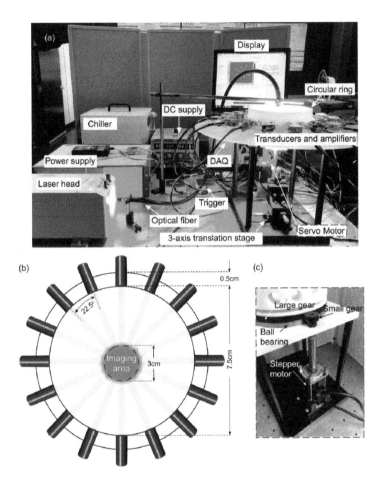

Figure 1. (**a**) A photograph of our low-cost photoacoustic computed tomography system comprised of EKSPLA laser head, a power supply for the laser, a chiller, a DC power supply for the motor driver, NI DAQ, NI trigger board, servo motor (expanded in (**c**)), motor gear, 3-axis translation stage for phantom, and 16 transducers and amplifiers, (**b**) top view of the circular ring showing an imaging area of 3 cm diameter, and (**c**) servo motor with mechanical gears. DC: Direct Current, NI: National Instrument, DAQ: Data Acquisition Unit.

The laser is coupled with 10mm diameter plastic PMMA optical fiber (epef-10, Ever Heng Optical Co., Shenzhen, China), for illumination of the imaging target. Output laser energy is ~32 mJ/cm^2; recorded using an energy meter (QE12SP-H-MT-D0, Gentec-EO, Quebec, QC, Canada). Considering the distance between the optical fiber and the sample, as well as divergence of light, we calculated the light energy onto the sample, ~19.1 mJ/cm^2, below the ANSI limit. The optical fiber is held in place by optical rods and positioned at the center of the scanner (circular ring) to illuminate the target with maximum uniformity. The energy loss in the optical fiber was ~30%.

The ultrasound detection unit consists of 16 single element 5 MHz ultrasound transducers (ISL-0504-GP, Technisonic Research Inc., Fairfield, CT, USA), inserted and fitted along the circumference of the circular ring made up of Polyactic Acid (PLA) plastic (15 cm diameter). These transducers are radially separated at 22.5° from one another as depicted in Figure 1. The object is separated from the water tank using a transparent saran wrap and the tank is filled with distilled water for ultrasound coupling as shown in Figure 2.

PA signals acquired from each transducer are amplified using a low-noise 24 dB amplifiers (ZFL-500LN, Mini-Circuits, Brooklyn, NY, USA). The amplified signals are then fed into a 16 channel DAQ (NI PXIe -1078, National Instruments, Austin, TX, USA). The NI system is equipped with two 14-bit, 8 channel data acquisition cards (NI PXIe-5170R, National Instruments, Austin, TX, USA). The data acquisition is synchronized by an internally generated trigger from the laser. The sampling rate is set to 50 MS/s. The number of samples in each PA signal acquired is 6000. Based on the pulse repetition rate of the laser and the number of channels, 1200 view angles are scanned in only 1.5 s; this data is then used for image reconstruction.

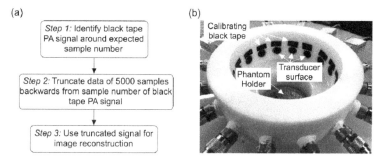

(a)

Step 1: Identify black tape PA signal around expected sample number

Step 2: Truncate data of 5000 samples backwards from sample number of black tape PA signal

Step 3: Use truncated signal for image reconstruction

(b)

Calibrating black tape

Phantom Holder

Transducer surface

Figure 2. (**a**) Flowchart of the correction technique and (**b**) calibration setup with black tape targets located above each transducer to acquire calibration PA signal for the alignment of the transducers' signals.

The ring is coupled to a servo motor (J0400-301-4-000, Applied Motion Products Inc., Watsonville, CA, USA) through mechanical gears (KHK SS1-200, KHK USA Inc., Mineola, NY, USA) for rotation of the system as shown in Figure 1c. Two meshed gears are used to rotate the circular ring. The small gear is mounted on a shaft that is directly connected to the servo motor and the large gear is mounted on a ball bearing with the same diameter of the circular ring to reduce friction and hence, increase the rotational speed. The small gear was chosen with higher pitch as compared to the large gear, further increasing rotational speed. With the objective of developing a fast scanning method, the gear system parameters were determined based on the weight of the system, moment of inertia, and holding torque of the motor as explained below.

For speed optimization, gear ratio was determined based on the ratio of the required torque to motor torque. The required torque was calculated based on the moment of inertia and angular motion that has an input criterion of 22.5° rotation per second. The moment of inertia, I, was calculated to be 0.0197 kg·m^2 using total mass and internal radius of the ring. If the motor rotates 22.5° in 1.5 s, then the angular rotation per minute, α, would be 24 and hence, the required torque, T_{reqd}, is 0.4728 N-m. According to the motor specification, motor torque, T_{motor}, is 0.056 N·m. Therefore, the gear ratio, η_g is rounded up to 9. Mass of water inside the ring is calculated considering a density of 1000 kg/m^3. Other structural parameters of the circular ring are provided in Table 1.

Table 1. Structural parameters of the circular ring.

Parameter	Value
Height, h	31 cm
External radius, r_{ext}	8 cm
Internal radius, r_{int}	7.5 cm
Motor weight, m_r	5 kg
Total weight, m_t	7.01 kg

The motor rotates 22.5° in 1.5 s. The data acquisition and motor rotation are synchronized using a LabView code. Based on the laser repetition rate, 75 view angles are defined within the 22.5°. The size of the acquired data is 1200 × 16 bits.

For reconstruction, we used filtered back projection algorithm [36–38]. The algorithm requires the scanning radius as an input. Unlike a commercial ring array, where the transducers have been perfectly manufactured to be equidistant from the center of the ring, in the proposed system, the transducers are not equidistant from the center due to the manual insertion of the transducers into the holes on the ring; i.e., if 7.5 cm is used as the radius of the ring for all the channels, the reconstruction will be imperfect and lead to a distorted image. In fact, there is a shift in the expected location of the PA signals coming from the imaging target. Therefore, the calibration method is required. The original values of radius are within 7.58–7.68 mm (please see Table 2). An easy fix will be to change the scanning radius until an image with adequate quality is obtained. This method, is however, time consuming and somewhat inefficient.

Table 2. The values of scanning radius for transducers.

Transducer no.	Distance to the Center (cm)
1	7.61
2	7.65
3	7.67
4	7.69
5	7.58
6	7.62
7	7.64
8	7.62
9	7.65
10	7.67
11	7.63
12	7.62
13	7.64
14	7.63
15	7.68
16	7.68

To address this issue and ensure all the transducers are equidistant from the center of the circular ring, a data correction algorithm is used. The flowchart of the correction methodology is provided in Figure 2a. In this method, we image high optical energy absorbent (trimmed homogenous black tape strips made of vinyl). A small piece of black tape is attached on the inner wall of the ring above each of the 16 transducers as shown in Figure 2. While imaging, the output laser from the optical fiber had sufficient energy to produce a distinguishable PA signal from the black tape for calibration. The PA signals generated from the black tape and imaging target do not overlap because they are spatially apart. Each transducer detects a PA signal from the tape that is located 180 degrees away from the transducer. The diameter of the ring, the distance between tape and the opposite transducer, is 15 cm. As a result, the PA signal generated from the tape is supposed to occur at the sample number corresponding to such distance, for all 16 channels.

Since the signal shift is the same for both PA signals coming from the tape and the imaging target, the number of samples between the peak of the PA signal from the tape and the imaging target remains the same. To align the data of each channel, the arrays are truncated to the same length (5000 samples) with the last sample being the PA signal of their respective calibration tape. Based on this methodology, a calibration algorithm is developed and applied on each channel data as shown in Figure 3.

In order to determine the system resolution, a 0.2 mm single lead phantom is made as shown in Figure 4a. This phantom is pivoted onto a white plastic platform using a separate 1.5 cm long lead. Alignment of the phantom with the plane of transducer field of view is performed by a 3-axis translation stage (Thorlabs, MT3 148-811ST). A complex 8-legged 0.5 mm diameter lead phantom is also developed to evaluate the system performance as shown in Figure 5a. The 8 legs of the phantom were held together using transparent non-conductive adhesive glue.

3. Results

Using the calibration algorithm, the PA channel data was corrected before image reconstruction. Four randomly selected channel data are shown in Figure 3. Before calibration, each channel data consists of three reference signals: (i) the signal representing the transducer response, (ii) the PA signal coming from the imaging target and (iii) the PA signal coming from the tape. After calibration (blue solid), the PA signal from the imaging target is aligned. The reason that the PA signals are not located at the same sample number, is due to the non-uniform structure of the phantom. The improvement in the reconstructed image is evident in Figure 3e,j.

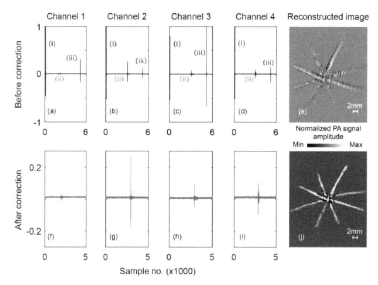

Figure 3. PA signal correction based on the calibration algorithm. (**a–d**) PA signals of (i) transducer response, (ii) PA from imaging object, (iii) PA from calibration tape, (**e**) distorted image before data correction. (**f–i**) PA signals from the imaging object, (**j**) image after data correction.

We initially performed a resolution analysis on a 2-legged phantom with 0.2 mm diameter lead as shown in Figure 4a. As the diameter of the lead is expected to be below the spatial resolution of the system, the width of the reconstructed image will spread to the resolution of the system. The length of each of the two branches was 6 mm. A 2D reconstructed image of the phantom is shown in Figure 4b. Taking a 1D intensity profile across the diameter of the object followed by a Gaussian fitting to the pixel values and calculating the full width at half maximum (FWHM) approximates the spatial resolution of the system. In Figure 4c, the FWHM is estimated to be 240 µm.

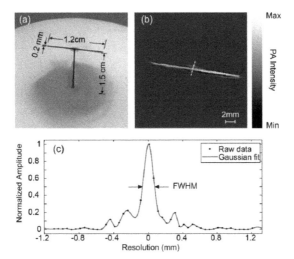

Figure 4. Resolution analysis results. (**a**) Image of the resolution phantom, (**b**) reconstructed image, (**c**) intensity profile across the diameter of the resolution phantom with a Gaussian fit (FWHM = 240 μm).

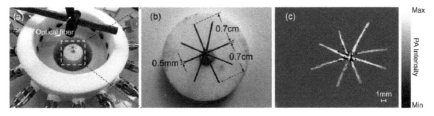

Figure 5. Imaging of an 8-leg phantom. (**a**) Experimental setup. (**b**) a photograph of the 8-leg phantom, and (**c**) a 2D reconstructed image of the phantom using 16- element PACT system (image acquired in 1.5 s).

Next, the system was used to image a complex phantom comprised of 0.5 mm diameter lead with 8 branches (~7 mm in length). Figure 5a shows an image of the phantom positioned and aligned at the center of the ring. The phantom dimensions are shown in Figure 5b. The reconstructed image of the phantom is shown in Figure 5c.

Among different configurations of PAI, photoacoustic computed tomography is favored for small animal studies, e.g., hemodynamic brain imaging, whole body imaging, mainly due to its fast image acquisition. However the high-cost of the system has prevented the full establishment and wide utility of the device. Here we introduced a fast, low cost PACT system by replacing the sophisticated expensive ring array with a moving ultrasound sparse array. We demonstrated, by imaging sophisticated phantoms, that the system can produce images with a temporal resolution of 1.5 s and spatial resolution of 240 μm. This system promises a higher sensitivity in comparison to a commercial ring array due to the much larger number of view angles it uses and larger element size of the transducers. The cost of the proposed system compared to a full-ring array PACT is given in the table below (Table 3).

Table 3. Cost comparison between proposed and full-ring array based PACT system.

Components	Proposed/Full-Ring Array	Cost ($)
Transducer	16 single elements/512 elements	~5K/~65K
Amplifiers	Low noise 24dB/customized	~5K/~25K
Servo Motor	Servo/NA	~250/NA
DAQ	16 channel/512 channel (or 64 channel with MUX)	~5K/~50 K

4. Conclusions and Future Work

Although PAI has demonstrated great potential in preclinical and clinical applications, it is still in its early stage of development compared to already established medical imaging modalities such as MRI, CT, PET, and ultrasound imaging. Among several limitations preventing the wide application of PACT, manufacturing cost is the major one. Here, we explained the development of a fast, low-cost PACT system with only 16 single element transducers and a novel mechanical scanning design. Such configuration reduces the cost of the ring array PACT system drastically. A correction algorithm has been developed and applied to the acquired signals to make the transducers' data equidistant from the imaging target. We have demonstrated that the system can produce images with a temporal resolution of 1.5 s and spatial resolution of 240 μm.

The system developed in this study will eventually be used for functional small animal brain imaging. We are currently developing a new system, using the same concept but smaller in size with 32 transducers, and an amplifier unit, to make a faster, compact PACT system with improved isotropic resolution.

Author Contributions: Conceptualization, M.Z. and M.A.; Methodology, M.Z. and M.A.; Software, M.Z., K.K., and R.M.; Validation, R.M. and K.K.; Formal Analysis, R.M., K.K., and M.Z.; Investigation R.M., K.K., and M.Z.; Writing–Original Draft Preparation, M.Z., and K.K.; Writing–Review and Editing, R.M.; Visualization, R.M.; Supervision, M.A.

Funding: This research was partially funded by the American Cancer Society, Research Grant number 14-238-04-IRG and the Albert and Goldye J. Nelson grant.

Acknowledgments: We are grateful to have constructive discussion with Jun Xia from University of Buffalo, NY regarding the reconstruction algorithm.

Conflicts of Interest: The authors declare no conflict of interest.

References

1. Wang, L.V. Multiscale photoacoustic microscopy and computed tomography. *Nat. Photonics* **2009**, *3*, 503. [CrossRef] [PubMed]
2. Mohammadi-Nejad, A.-R.; Mahmoudzadeh, M.; Hassanpour, M.S.; Wallois, F.; Muzik, O.; Papadelis, C.; Hansen, A.; Soltanian-Zadeh, H.; Gelovani, J.; Nasiriavanaki, M. Neonatal brain resting-state functional connectivity imaging modalities. *Photoacoustics* **2018**, *10*, 1–19. [CrossRef] [PubMed]
3. Anwar, R.; Kratkiewicz, K.; Mohammad, R. Avanaki Photoacoustic Imaging: A Promising Alternative to Transcranial Ultrasound. *Res. J. Opt. Photonics* **2018**, *2*, 411–420.
4. Wang, L.V. Tutorial on photoacoustic microscopy and computed tomography. *IEEE J. Sel. Top. Quantum Electron.* **2008**, *14*, 171–179. [CrossRef]
5. Zhou, Y.; Yao, J.; Wang, L.V. Tutorial on photoacoustic tomography. *J. Biomed. Opt.* **2016**, *21*, 061007. [CrossRef] [PubMed]
6. Nasiriavanaki, M.; Xia, J.; Wan, H.; Bauer, A.Q.; Culver, J.P.; Wang, L.V. High-resolution photoacoustic tomography of resting-state functional connectivity in the mouse brain. *Proc. Natl. Acad. Sci. USA* **2014**, *111*, 21–26. [CrossRef] [PubMed]
7. Mahmoodkalayeh, S.; Lu, X.; Ansari, M.A.; Li, H.; Nasiriavanaki, M. Optimization of light illumination for photoacoustic computed tomography of human infant brain. In Proceedings of the Photons Plus Ultrasound: Imaging and Sensing, San Francisco, CA, USA, 27 January–1 February 2018; p. 104946U.

8. Manwar, R.; Hosseinzadeh, M.; Hariri, A.; Kratkiewicz, K.; Noei, S.; N Avanaki, M. Photoacoustic Signal Enhancement: Towards Utilization of Low Energy Laser Diodes in Real-Time Photoacoustic Imaging. *Sensors* **2018**, *18*, 3498. [CrossRef] [PubMed]

9. Xu, M.; Wang, L.V. Photoacoustic imaging in biomedicine. *Rev. Sci. Instrum.* **2006**, *77*, 041101. [CrossRef]

10. Zhou, Y.; Wang, D.; Zhang, Y.; Chitgupi, U.; Geng, J.; Wang, Y.; Zhang, Y.; Cook, T.R.; Xia, J.; Lovell, J.F. A phosphorus phthalocyanine formulation with intense absorbance at 1000 nm for deep optical imaging. *Theranostics* **2016**, *6*, 688. [CrossRef] [PubMed]

11. Adabi, S.; Hosseinzadeh, M.; Noei, S.; Conforto, S.; Daveluy, S.; Clayton, A.; Mehregan, D.; Nasiriavanaki, M. Universal in vivo textural model for human skin based on optical coherence tomograms. *Sci. Rep.* **2017**, *7*, 17912. [CrossRef] [PubMed]

12. Choma, M.A.; Sarunic, M.V.; Yang, C.; Izatt, J.A. Sensitivity advantage of swept source and Fourier domain optical coherence tomography. *Opt. Express* **2003**, *11*, 2183–2189. [CrossRef] [PubMed]

13. Cogliati, A.; Canavesi, C.; Hayes, A.; Tankam, P.; Duma, V.-F.; Santhanam, A.; Thompson, K.P.; Rolland, J.P. MEMS-based handheld scanning probe with pre-shaped input signals for distortion-free images in Gabor-domain optical coherence microscopy. *Opt. Express* **2016**, *24*, 13365–13374. [CrossRef] [PubMed]

14. Monroy, G.L.; Won, J.; Spillman, D.R.; Dsouza, R.; Boppart, S.A. Clinical translation of handheld optical coherence tomography: practical considerations and recent advancements. *J. Biomed. Opt.* **2017**, *22*, 121715. [CrossRef] [PubMed]

15. Liu, Y.; Nie, L.; Chen, X. Photoacoustic molecular imaging: from multiscale biomedical applications towards early-stage theranostics. *Trends Biotechnol.* **2016**, *34*, 420–433. [CrossRef] [PubMed]

16. Hariri, A.; Tavakoli, E.; Adabi, S.; Gelovani, J.; Avanaki, M.R. Functional photoacoustic tomography for neonatal brain imaging: developments and challenges. In Proceedings of the Photons Plus Ultrasound: Imaging and Sensing, San Francisco, CA, USA, 28 January–2 February 2017; p. 100642Z.

17. Mahmoodkalayeh, S.; Jooya, H.Z.; Hariri, A.; Zhou, Y.; Xu, Q.; Ansari, M.A.; Avanaki, M.R. Low temperature-mediated enhancement of photoacoustic imaging depth. *Sci. Rep.* **2018**, *8*, 4873. [CrossRef] [PubMed]

18. Meimani, N.; Abani, N.; Gelovani, J.; Avanaki, M.R. A numerical analysis of a semi-dry coupling configuration in photoacoustic computed tomography for infant brain imaging. *Photoacoustics* **2017**, *7*, 27–35. [CrossRef] [PubMed]

19. Wang, L.V.; Yao, J. A practical guide to photoacoustic tomography in the life sciences. *Nat. Methods* **2016**, *13*, 627. [CrossRef] [PubMed]

20. Laufer, J.; Delpy, D.; Elwell, C.; Beard, P. Quantitative spatially resolved measurement of tissue chromophore concentrations using photoacoustic spectroscopy: application to the measurement of blood oxygenation and haemoglobin concentration. *Phys. Med. Biol.* **2006**, *52*, 141. [CrossRef] [PubMed]

21. Stein, E.W.; Maslov, K.I.; Wang, L.V. Noninvasive, in vivo imaging of blood-oxygenation dynamics within the mouse brain using photoacoustic microscopy. *J. Biomed. Opt.* **2009**, *14*, 020502. [CrossRef] [PubMed]

22. Wang, X.; Xie, X.; Ku, G.; Wang, L.V.; Stoica, G. Noninvasive imaging of hemoglobin concentration and oxygenation in the rat brain using high-resolution photoacoustic tomography. *J. Biomed. Opt.* **2006**, *11*, 024015. [CrossRef] [PubMed]

23. Wang, L.V.; Gao, L. Photoacoustic microscopy and computed tomography: from bench to bedside. *Annu. Rev. Biomed. Eng.* **2014**, *16*, 155–185. [CrossRef] [PubMed]

24. Wang, L.V.; Hu, S. Photoacoustic tomography: in vivo imaging from organelles to organs. *Science* **2012**, *335*, 1458–1462. [CrossRef] [PubMed]

25. Upputuri, P.K.; Pramanik, M. Performance characterization of low-cost, high-speed, portable pulsed laser diode photoacoustic tomography (PLD-PAT) system. *Biomed. Opt. Express* **2015**, *6*, 4118–4129. [CrossRef] [PubMed]

26. Xia, J.; Chatni, M.R.; Maslov, K.; Guo, Z.; Wang, K.; Anastasio, M.; Wang, L.V. Whole-body ring-shaped confocal photoacoustic computed tomography of small animals in vivo. *J. Biomed. Opt.* **2012**, *17*, 0505061–0505063. [CrossRef] [PubMed]

27. Xia, J.; Guo, Z.; Maslov, K.; Aguirre, A.; Zhu, Q.; Percival, C.; Wang, L.V. Three-dimensional photoacoustic tomography based on the focal-line concept. *J. Biomed. Opt.* **2011**, *16*, 090505. [CrossRef] [PubMed]

28. Gamelin, J.K.; Aquirre, A.; Maurudis, A.; Huang, F.; Castillo, D.; Wang, L.V.; Zhu, Q. Curved array photoacoustic tomographic system for small animal imaging. *J. Biomed. Opt.* **2008**, *13*, 024007. [CrossRef] [PubMed]

29. Buxton, R.B.; Wong, E.C.; Frank, L.R. Dynamics of blood flow and oxygenation changes during brain activation: the balloon model. *Magn. Reson. Med.* **1998**, *39*, 855–864. [CrossRef] [PubMed]

30. Gamelin, J.; Maurudis, A.; Aguirre, A.; Huang, F.; Guo, P.; Wang, L.V.; Zhu, Q. A fast 512-element ring array photoacoustic imaging system for small animals. In Proceedings of the Photons Plus Ultrasound: Imaging and Sensing, San Jose, CA, USA, 24–29 January 2009; p. 71770B.

31. Li, G.; Xia, J.; Wang, K.; Maslov, K.; Anastasio, M.A.; Wang, L.V. Tripling the detection view of high-frequency linear-array-based photoacoustic computed tomography by using two planar acoustic reflectors. *Quant. Imaging Med. Surg.* **2015**, *5*, 57. [PubMed]

32. Kondo, K.; Namita, T.; Yamakawa, M.; Shiina, T. Three-dimensional photoacoustic reconstruction for sparse array using compressed sensing based on k-space algorithm. In Proceedings of the IEEE International Conference on the Ultrasonics Symposium (IUS), Tours, France, 18–21 September 2016; pp. 1–3.

33. Xiao, J.; He, J. Multispectral quantitative photoacoustic imaging of osteoarthritis in finger joints. *Appl. Opt.* **2010**, *49*, 5721–5727. [CrossRef] [PubMed]

34. Wang, X.; Pang, Y.; Ku, G.; Xie, X.; Stoica, G.; Wang, L.V. Noninvasive laser-induced photoacoustic tomography for structural and functional in vivo imaging of the brain. *Nat. Biotechnol.* **2003**, *21*, 803. [CrossRef] [PubMed]

35. Liu, Y.; Li, D.; Yuan, Z. Photoacoustic tomography imaging of the adult zebrafish by using unfocused and focused high-frequency ultrasound transducers. *Appl. Sci.* **2016**, *6*, 392. [CrossRef]

36. Xu, M.; Wang, L.V. Time-domain reconstruction for thermoacoustic tomography in a spherical geometry. *IEEE Trans. Med. Imaging* **2002**, *21*, 814–822. [PubMed]

37. Xu, M.; Wang, L.V. Universal back-projection algorithm for photoacoustic computed tomography. *Phys. Rev. E* **2005**, *71*, 016706. [CrossRef] [PubMed]

38. Omidi, P.; Zafar, M.; Mozaffarzadeh, M.; Hariri, A.; Haung, X.; Orooji, M.; Nasiriavanaki, M. A novel dictionary-based image reconstruction for photoacoustic computed tomography. *Appl. Sci.* **2018**, *8*, 1570. [CrossRef]

Article

Adipocyte Size Evaluation Based on Photoacoustic Spectral Analysis Combined with Deep Learning Method

Xiang Ma [1], Meng Cao [1], Qinghong Shen [1], Jie Yuan [1,*], Ting Feng [1,2], Qian Cheng [2], Xueding Wang [2,3], Alexandra R. Washabaugh [4], Nicki A. Baker [4], Carey N. Lumeng [5] and Robert W. O'Rourke [4,6]

[1] Schoool of Electronic Science and Engineering, Nanjing University, Xianlin, Nanjing 210046, China; xiangma@smail.nju.edu.cn (X.M.); njucaomeng@163.com (M.C.); qhshen@nju.edu.cn (Q.S.); fengting@njust.edu.cn (T.F.)
[2] Institute of Acoustics, Tongji University, Shanghai 200092, China; q.cheng@tongji.edu.cn (Q.C.); xdwang@umich.edu (X.W.)
[3] Department of Biomedical Engineering, University of Michigan, Ann Arbor, MI 48109, USA
[4] Department of Surgery, University of Michigan Medical School and Michigan Medicine, Ann Arbor, MI 48109, USA; alexwash@med.umich.edu (A.R.W.); nickiba@med.umich.edu (N.A.B.); rorourke@med.umich.edu (R.W.O.)
[5] Department of Pediatrics and Communicable Diseases, University of Michigan Medical School and Michigan Medicine, Ann Arbor, MI 48109, USA; clumeng@med.umich.edu
[6] Ann Arbor Veterans Administration Hospital, Ann Arbor, MI 48109, USA
*   Correspondence: yuanjie@nju.edu.cn

Received: 10 October 2018; Accepted: 5 November 2018; Published: 7 November 2018

Abstract: Adipocyte size, i.e., the cell area of adipose tissue, is correlated directly with metabolic disease risk in obese humans. This study proposes an approach of processing the photoacoustic (PA) signal power spectrum using a deep learning method to evaluate adipocyte size in human adipose tissue. This approach has the potential to provide noninvasive assessment of adipose tissue dysfunction, replacing traditional invasive methods of evaluating adipose tissue via biopsy and histopathology. A deep neural network with fully connected layers was used to fit the relationship between PA spectrum and average adipocyte size. Experiments on human adipose tissue specimens were performed, and the optimal parameters of the deep learning method were applied to establish the relationship between the PA spectrum and average adipocyte size. By studying different spectral bands in the entire spectral range using the deep network, a spectral band mostly sensitive to the adipocyte size was identified. A method of combining all frequency components of PA spectrum was tested to achieve a more accurate evaluation.

Keywords: photoacoustics; tissue characterization; absorption

1. Introduction

Obesity is a public health crisis which afflicts one third of the U.S. population [1] and is associated with a significant risk of metabolic diseases. Metabolic diseases come with abnormal chemical reactions in the body; diabetes mellitus (DM) is one of the most common metabolic diseases. The National Health and Nutrition Examination Survey (NHANES) reveals that diabetes is becoming increasingly prevalent and the corresponding weight classes are continually increasing, with nearly half of adult diabetics considered to be obese [2]. Prediction of obese patients at risk for developing metabolic disease is of central importance. Prior research demonstrates that adipose tissue phenotypic features, also known as tissue observable characteristics, such as adipocyte size, correlate with human metabolic disease [3–5]. That suggests the adipose size may serve as predictors of metabolic disease risk and a diagnostic and therapeutic target. Histology is the current gold standard for assessments of adipose size [3–5]. While accurate, histology requires adipose tissue biopsy, which involves surgery and microscope observation, which is invasive and labor- and cost-intensive.

Emerging biomedical photoacoustic (PA) imaging technology is based upon the detection of laser-induced thermoacoustic signals from a biological sample which can then be used to produce images reflecting the optical absorption contrast in the sample [6–8]. Due to PA signal's broad-band feature, it can reflect sample microstructures of different scopes. Recently, various methods for analyzing the PA signals in the frequency domain have been developed by several research groups [9–12], aiming to achieve quantitative evaluation of tissue microstructures. One of the methods termed as photoacoustic spectral analysis (PASA) [13–16], borrowed the basic concept from ultrasound spectral analysis [17–19]. In PASA, by quantifying the key parameters, including slope, intercept, and mid-band fit, of the linear fit to the truncated signal power spectrum within the predetermined frequency range, histological microfeatures of the optically absorbing materials in target tissues were characterized. Later, by using ultrabroad-band ultrasonic detectors, the scope of PASA was further extended from the tissue level to the cellular level [16]. Using the three parameters of the linear fit (i.e., slope, intercept, and mid-band fit) to describe the main features of the power spectrum provides a practical way of addressing the stochastic nature of the tissue microstructure, and can lead to measurements that are not only quantitative but also robust. However, despite these advantages, PASA, relying on a simple linear fit to the power spectrum within the entire predetermined frequency range, utilizes only limited information of the signal power spectrum.

Spectral analysis based on deep learning provides a potential solution to this problem. A network is established in deep learning which describes the relationships between input and output by parameters and functions. The process of training uses the data sets whose input and output are known to modify all the parameters in the network to achieve an optimal model under control of a learning rule [20]. Upon achieving optimal parameters, the network can be used to predict the corresponding output of data sets whose input is known. Varying layer numbers (numbers of processing steps) and layer sizes (numbers of parameters involved in each processing step) can provide different performance of data representation which results in varying accuracy of prediction [21]. Due to its excellent performance characteristics, such as high accuracy, deep learning methods in medical diagnosis and evaluation have been rapidly developed [22–24].

We studied the feasibility of measuring adipocyte size in human adipose tissue by analyzing PA spectrum with the help of deep learning in this research. A deep neural network was used to fit the relationship between the PA spectrum and an average adipocyte. We hypothesized that PA measurement powered by deep learning can provide accurate assessment of adipocyte size as determined by the gold standard histology. To examine the validity of this hypothesis, experiments on *ex vivo* adipose tissue specimens from diabetic and nondiabetic obese human subjects were conducted and the network was trained in different spectral sections to obtain the results.

In this paper, we first present the principle of using deep network to evaluate the average adipocyte size. And then we show the procedure of *ex vivo* experiment and details of building our data set. Next, results of using networks with different layer numbers are compared and experiments of analyzing different spectral bands and the entire spectrum are presented. Finally, we make some comparisons between our deep learning method and traditional PASA method.

2. Deep Neural Network with Fully Connected Layers

The traditional PASA method only depends on a simple linear fit with the power spectrum. Therefore, it cannot accurately represent the relationship between PA spectrum and average adipocyte size which may be complex and difficult to be fit with an intelligible mathematical expression. Compared to PASA, deep learning methods can provide a much more complicated nonlinear model to fit the relationship and achieve higher accuracy for our study. More specifically, with each A-line PA signal from a tissue, the power spectrum was computed and normalized, and then analyzed using deep learning method to fully utilize the spectral information. With the method of deep learning, the relationship between the power spectrum and adipocyte size was learned by a deep network. The power spectrum and the corresponding adipocyte size worked as the input and output of the deep network, respectively, when training and testing.

For an N-layer network, the element number of input and output of the network are I_0 and O_{N+1}. In our study, each element of the input was a power spectrum value of the PA signal while the output, with only one element, was the average adipocyte size. With the power spectrum of a PA signal fed into the first layer, all output elements of each layer functioned as the input of the next layer and the output of the last layer was the predicted average adipocyte size.

In our study, N-1 hidden layers and one output layer (the last layer) were used to fit the nonlinear relationship between power spectrum and average adipocyte size. Each layer is a fully connected layer, which means all of the input elements are connected to each output element, as shown in Figure 1. All hidden layers and output layer did a linear transformation at first and then used different functions to map linear transformation result to output. The n-th layer can be defined as

$$y_j = f\left(\sum_{i=0}^{I_n} w_{i,j} x_i + b_j\right), j = 1, 2, \ldots, O_n, \tag{1}$$

where x is the input of the layer (I_n elements) and y is the output (O_n elements). The transfer function $f(x)$, as shown in Figure 1b, in hidden layers was selected as tan-sigmoid function in our network, which is defined as

$$f(x) = \frac{2}{1 + e^{-2x}} - 1, x \in (-\infty, +\infty), \tag{2}$$

The tan-sigmoid function is continuous, smooth, and monotone, and increases quickly in the vicinity of $x = 0$ with a codomain of $(-1, 1)$. Here, we chose a tan-sigmoid function as the transfer function because its features match with the human adipocyte size distribution, as their values should be continuous and bounded. For output layer, transfer function $f(x)$ was selected as a linear function so that the output can match the adipocyte size appropriately.

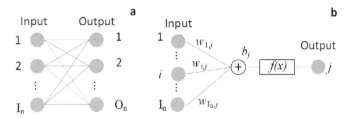

Figure 1. One layer of deep network. (**a**) Fully connected layer. All of the input elements are connected to each output element. (**b**) Obtaining j-th element of output from input elements.

Through setting layer number N and the input and output element number for each layer, we establish a network and use all parameters (w and b in Equation (1)) to fit the relationship between input and output. When training, data set was divided into training set and test set. The training set was randomly picked out from the whole data set while the remains were defined as test set. All parameters (w and b in Equation (1)) in the network were optimized gradually through iteration to minimize the difference between the predicted average adipocyte size and the ground truth value from histology for all samples in the training set. When testing, the power spectrum of test set was fed into the network with the parameters obtained through training to predict the average adipocyte sizes. When calculating the error, we first calculated the relative error δx of every sample from test set, which can be defined as

$$\delta x = \frac{\Delta x}{x} = \frac{|x_0 - x|}{x}, \tag{3}$$

where Δx is absolute error. It is equal to the absolute difference between the output of the deep learning network x_0 and the ground truth value x from histologic analysis. Next, the average of absolute errors and the average of relative errors of all samples from the test set were calculated to get the MAE (mean absolute error) and MRE (mean relative error).

3. Results and Discussions

3.1. Data Set Building

An experiment on *ex vivo* human adipose tissue specimens was conducted to test our method. To build our data set for deep learning method, we studied the histological photographs of visceral adipose tissue samples collected from 28 human subjects (12 diabetic subjects and 16 nondiabetic subjects) and calculated the accurate average size for each sample. Human subjects undergoing bariatric surgery were enrolled with Institutional Review Board approval from the University of Michigan and Ann Arbor Veterans Administration Hospital. Visceral adipose tissue (VAT) from the greater omentum was collected at the beginning of operation and processed immediately. By cutting collected tissues into small specimens we obtained 110 specimens of adipose. Then each specimen was divided into two parts which were used to for microscope analysis and to acquire PA signal, respectively. Adipose tissue explants were fixed in 10% formalin, embedded in paraffin, and sectioned onto charged glass slides. Fixed hematoxylin/eosin-stained slides were imaged on an microscope and analyzed with ImageJ software by a blinded observer, as previously described [1]. An example of histological photograph is shown in Figure 2c. For each histological photograph, pixel areas of all individual cells were averaged to evaluate the adipocyte size and the size distribution of the sample in Figure 2c and its Gamma fitting, which proved to be the best fit of adipocyte size distribution in previous research [17], are shown in Figure 2d.

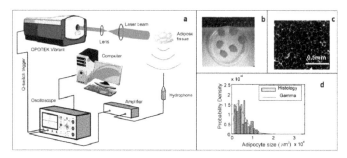

Figure 2. Experiment on *ex vivo* human adipose tissue specimens. (**a**) Schematic diagram of experimental setup for PA signal acquisition. (**b**) Photograph of human adipose tissue specimens. (**c**) Representative microphotograph of a human adipose tissue sample demonstrating adipocyte microarchitecture. (**d**) Adipocyte size distribution corresponding to (**c**).

PA signal acquisition was conducted by using the setup shown in Figure 2a. Laser light was from a tunable Optical Parametric Oscillator pumped by the second harmonic output of an Nd:YAG pulsed laser. The laser illuminated the samples at the wavelength of 1210 nm where lipid has strong optical absorption [25]. During signal acquisition, the laser illuminated the whole adipose tissue specimen, achieving an averaged light fluence of 5.6 mJ/cm^2 on the sample surface. Each specimen was cut into an ellipsoid-like shape; sizes shown in Table 1. A needle hydrophone was set against the samples to acquire PA signals. Through a preamplifier and a low-noise amplifier, the PA signal received by the hydrophone was recorded with a digital oscilloscope. When recording PA signal, averaging was done by oscilloscope for 50 times for one PA signal. We listed the experiment parameters in Table 1.

Table 1. Experiment parameter setting.

Parameter	Value
Sample number	110 in total, 48 from diabetic subjects, 62 from nondiabetic subjects
Sample size for PA signal acquisition	Approximately 15 mm diameter 5–6 mm thickness
Microscope image	Olympus IX-81 fluorescent microscope, captured as multiple TIFF-gray-scale images
ImageJ	Version 1.42, National Institutes of Health and University of Wisconsin, Madison, WI, USA
Optical Parametric Oscillator	OPO, Vibrant 532 I, Opotek, Carlsbad, CA, USA
Nd:YAG pulsed laser	Brilliant B, Quantel, Bozeman, MT, USA
Laser beam	1210-nm wavelength, 1.5-cm diameter, 10-mJ pulse energy, 5-ns pulse duration
Needle hydrophone	5 cm, 0.5 mm^2, NIH Resource Center for Medical Ultrasonic Transducer Technology, University of Southern California, United States, detection bandwidth of 52.5 MHz was centered at 35 MHz
Preamplifier	18 dB ZFL-1000+, Mini-Circuits, Brooklyn, NY, USA
Amplifier	40 dB, 5072PR, Parametrics, Waltham, MA, USA
Oscilloscope	TDS 540, Tektronix, Beaverton, OR, USA
Sampling rate	250 MS/s

Considering the limitation of our sample size, we also performed data augmentation [26] to enlarge data set size. Before each training, we first randomly picked out 90% of the 110 samples (i.e., 99 samples) as a training set and the left (i.e., 11 samples) as a test set. Next, as shown in Figure 3a, the signal of each sample was divided into three parts that shared the same length and cover effective signal so that 297 samples and 33 samples were obtained for training set and test set respectively [27]. We also added white Gaussian noise to signals of training set and another two signals were obtained for each signal so that totally 891 samples were used for training. We assumed that the average adipocyte size for each signal was equal to the average adipocyte size of the signal which generated it. In our study, we performed this data augmentation each time before training.

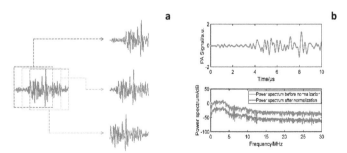

Figure 3. Data augmentation. (**a**) Raw signal was divided into three parts based on time. (**b**) Signal from (**a**) with noise and its power spectrum in log scale before and after normalization.

The power spectrum of each signal obtained through data augmentation was calculated. To reduce the effects of laser energy fluctuation on the signal, normalization was performed in the frequency domain by making magnitudes of all frequency components divided by their sum, which can be defined as

$$P_{norm}(k) = \frac{P(k)}{\sum\limits_{i=0}^{N-1} P(i)}, k = 0, 1, \ldots, N-1, \tag{4}$$

where $P(k)$ is the power spectrum of the PA signal and $P_{norm}(k)$ is the power spectrum after normalization. One example of power spectrum in log scale are shown in Figure 3b.

In this study on human adipose tissue specimens, we focused on the spectral band of 0.5 to 24.5 MHz. The minimum value of 0.5 MHz was chosen to remove the low frequency noise below 0.5 MHz which was mainly caused by the laser illumination on the surface of the hydrophone. The maximum value of 24.5 MHz was chosen because the power spectrum before normalization reached the noise level (-40 dB) at this frequency. We aimed to test the validity and potential of deep learning method for adipocyte size evaluation based on the dataset.

3.2. Most Sensitive Spectral Band

We first tried to find out the most sensitive spectral band which can lead to accurate evaluation of average adipocyte size. Based on features of our data set, we predicted the spectral band of 12.5 to 16.5 MHz as the most sensitive. PA signals are wide-band signals and can reflect histological microfeatures of adipocyte tissues. The average adipocyte size of our data set was mainly within the range of 3600 to 6400 μm². As depicted in Figure 2c, an adipocyte is shaped like a polygon. If we consider the adipocyte as a square, the relationship between its diagonal length l and area (i.e., size) S can be expressed as the equation $S = l^2/2$. Thus, the average diagonal length of our data set mainly ranged from 84.85 μm to 113.14 μm. The corresponding time t it takes for sound to pass through adipocyte along diagonal can be computed, using the equation $t = l/c_0$, where c_0 is the speed of sound considered as 1450 m/s (a typical value in human adipose tissue [28]). Here, we used diagonal to estimate t because it is the longest path for sound to pass through, which varies when size changes. Therefore, the most sensitive frequency of our data set is considered to be

$$f = \frac{1}{t} = \frac{c_0}{d} \tag{5}$$

which ranged from 12.82 MHz to 17.09 MHz. Therefore, we first tested our deep learning method on the spectral band of 12.5 to 16.5 MHz.

We picked out the normalized magnitude of spectral band of 12.5 to 16.5 MHz as the input of the deep network; the corresponding average adipocyte size acquired from standard histology of the tissue worked as the output. The input had 80 elements, each correspondent to the magnitude over a 0.05-MHz step in the 4-MHz spectral band. To initialize and train the deep network, a MATLAB (2017, MathWorks, Natick, MA, USA) toolbox called Neural Network Toolbox was employed. When training, the Levenberg–Marquardt method [29] was selected as the optimization algorithm. As described in Section 2, the deep network had several layers and each layer was a full connected layer shown in Figure 1. The output of each layer works as the input of the next layer. Different numbers of layers can make up different networks which may show various performance. We tried on networks with different layer numbers to find the optimal layer number. For each layer number, the element number of the output of the first hidden layer was set to about half of the input element number to reduce the number of parameters in the network and for rest layers, the element number of the output decreased gradually. The network was retrained for a total of 10 times for each layer number. For each time, data augment, described in Section 3.2, was performed and the minimum, maximum, and average values of the 10 MREs of the test set were computed, the result of which is shown in Figure 4.

The depth (number of layers) of the deep network largely influences the performance of the network, as illustrated in Figure 4. Initially when the layer number increases, the network becomes deeper and a more complicated model is established to process the sample data, as a result of which, the MRE of the test set becomes smaller and the deep network is more stable as demonstrated by the smaller deviation range. However, additional layers mean more parameters of the deep network and the longer training time. In addition, problems such as vanishing gradient problem of tan-sigmoid function, overfitting, and the stronger influence of the noise on the network may appear when the network becomes too deep [30]. Therefore, when the layer number is larger than seven, further increasing the layer number leads to larger MRE and larger deviation range reflecting unstable performance. As a result, a 7-layer network turned out to be the best choice.

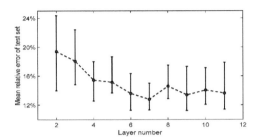

Figure 4. Training and testing using the spectral band of approximately 12.5 to 16.5 MHz with different layer numbers. The minimum, maximum, and average of MREs of test set for 10 retrainings are drawn for each layer number.

With the optimal network of seven layers which is shown in Figure 5, we also performed deep learning method on other bands of the spectral range with the same data set to test whether spectral band of 12.5 to 16.5 MHz really leads to the most accurate evaluation of average adipocyte size among the different spectral bands. The spectral range of 0.5 to 24.5 MHz was divided into twenty-one spectral bands. Each spectral band had a length of 4 MHz as well as a 3 MHz (75%) overlap with the adjacent spectral band. For each of the twenty-one spectral bands, the normalized magnitude of the 4-MHz spectral band worked as the input of the deep network, while the corresponding average adipocyte size worked as the output. Using data augmentation illustrated in Section 3.2, the deep network was trained for each band. Then, the relationship found out from the training was tested using the test set. For each of the 4-MHz spectral bands, the network was retrained for 10 times, and the average error of each time was computed. The average MRE from the 10 retrainings over different spectral bands are listed in Table 2.

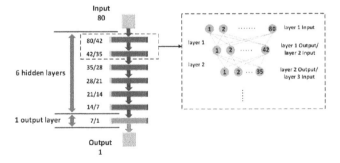

Figure 5. The structure of deep network. The numbers of the elements of the input and the output are marked on the left of each layer in the form of "I_m/O_m" for the *m*-th layer. All the layers are connected in the same as shown in Figure 1. The details of layer 1 and layer 2 are presented in the right as an example.

Table 2. Errors from the test set over different spectral bands.

Spectral Band/MHz	Average of MRE	Spectral Band/MHz	Average of MRE
0.5~4.5	19.04%	11.5~15.5	13.27%
1.5~5.5	18.59%	**12.5~16.5**	**12.78%**
2.5~6.5	18.53%	13.5~17.5	13.93%
3.5~7.5	16.48%	14.5~18.5	14.50%
4.5~8.5	16.73%	15.5~19.5	14.16%
5.5~9.5	14.82%	16.5~20.5	15.15%
6.5~10.5	14.10%	17.5~21.5	16.41%
7.5~11.5	14.41%	18.5~22.5	16.78%
8.5~12.5	13.85%	19.5~23.5	16.02%
9.5~13.5	13.79%	20.5~24.5	16.10%
10.5~14.5	13.92%		

By studying each 4-MHz spectral band in the entire spectral band using the deep network, the spectral band of 12.5 to 16.5 MHz turned out to be mostly sensitive to the adipocyte size. As is shown in Figure 6, in the range of 0.5 to 16.5 MHz, spectral bands covering higher frequencies lead to smaller errors; while in the range of 16.5 to 24.5 MHz, the error becomes larger when the frequency is higher. In addition to changes in MREs, the range of errors also changes with the spectral bands as well. In the spectral band of 0.5 to 4.5 MHz, the error varies in a large range. The range of errors decreases when the spectral band moves from 0.5 to 4.5 MHz to 12.5 to 16.5 MHz. However, when the spectral bands changes from 12.5 to 16.5 MHz to 20.5 to 24.5 MHz, the range of errors also increases. As a result, the spectral range of 12.5 to 16.5 MHz leads to the minimum error, and also a much smaller range of errors. The variation of error in different spectral bands results from the adipocyte size distribution. As illustrated in Figure 2d, we used Gamma distribution to fit the adipocyte size distribution. Number of adipocytes with sizes close to average size is larger than that of adipocytes whose size is away from the average. Therefore, the spectral band which is closer to the most sensitive spectral band corresponds to adipocytes accounting for a larger proportion and makes the performance of deep learning network more accurate and stable.

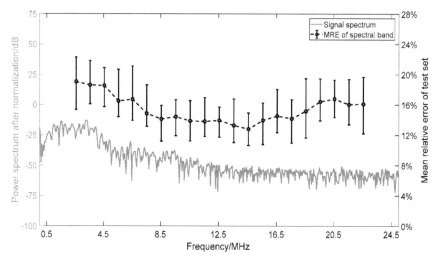

Figure 6. Plot showing mean relative error (MRE) of test set and the corresponding spectral bands. For each spectral band, the minimum, maximum and average of MREs of test set for 10 retrainings are drawn at the center frequency.

3.3. Training on the Entire Spectral Band

In addition to the most sensitive spectral band, we also tested our network on the entire spectral band with the same data set as we used on the most sensitive spectral band and we assumed that using all frequency components could contribute to a smaller error since PA signal is broad-band signal. The lower spectral range of 0.5 to 10.5 MHz contains information of outlines of adipocytes in tissue samples. It shows the size of adipocyte group which consists of several adjacent adipocytes. Sizes of these adipocyte groups are relative to numbers and sizes of each individual adipocyte. Indicated by Equation (5), a lower frequency is correspondent to a larger size of adipocyte group which consists of more or larger adjacent adipocytes. Thus, the spectral bands in this range contribute to evaluation of average adipocyte size but with a larger MRE (more than 14%) as listed in Table 2. Higher frequency in spectral range of 10.5 to 24.5 MHz is correspondent to different adipocyte sizes estimated by Equation (5). A higher frequency is relative to a smaller adipocyte size. Therefore, different frequency components indicate adipocyte size distribution and combining spectral information of all frequency components may lead to a smaller error.

To examine our hypothesis, our network was trained on the entire spectral band to obtain the error of combining all spectral components. Based on the normalized magnitude over a 0.05-MHz step in the range of 0.5 to 24.5 MHz, calculated when testing the network on different spectral bands, we calculated the sum of every six consecutive magnitude values and obtained the normalized magnitude over a 0.3-MHz step so that 80 input elements covered the entire spectral range of 0.5 to 24.5 MHz. Then, with normalized magnitude of the entire spectral band of 0.5 to 24.5 MHz with a 0.3-MHz step as the input and the correspondent average adipocyte size as the output, a network with the structure shown in Figure 5 was retrained 10 times. We performed the data augmentation for each retraining. The error distribution of all tested samples (330 samples) from the 10 retrainings is shown in Figure 7. As listed in Table 3, an average MAE of 394.19 μm² and MRE of 9.30% was achieved, which is smaller than errors of using different spectral bands listed in Table 2. The result indicated that combining all frequency components could lead to more accurate evaluations.

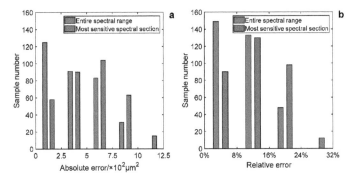

Figure 7. Error distribution of all tested samples of 10 retrainings using the most sensitive spectral band (12.5–16.5 MHz) and the entire spectral band (0.5–24.5 MHz). (**a**) Absolute error distribution for all tested samples. Absolute errors are mainly within the range of 0 to 500 μm² when training on the entire spectral band. The distribution of the relative errors is shown in (**b**).

Table 3. Mean absolute errors (MAEs) and MREs of the test set in 10 retrainings using the most sensitive spectral band and entire band.

Spectral Section	Error	Minimum	Maximum	Average
Entire spectral band (0.5~24.5 MHz)	MAE/μm²	326.76	455.84	394.19
	MRE	7.28%	10.94 %	9.30 %
Most sensitive spectral band (12.5~16.5 MHz)	MAE/μm²	463.61	609.06	548.37
	MRE	10.62%	14.90%	12.78%

3.4. Training on the Entire Spectral Band

We also compared the performance of the deep learning method with the conventional PASA method relying on linear fitting over the entire spectral band. This conventional method was used in a previous study to process the same data set from the human adipose tissue specimens [17]. With the power spectrum obtained from each tissue specimen over a spectral range of 0.5 to 24.5 MHz, a linear fit was produced which led to a quantified spectral parameter of *slope*. We assume that the relationship between the *slope* and the average adipocyte size follows a polynomial regression model of $y = ax^{-3} + bx^{-2} + cx^{-1} + d$, where y is *slope* and x is the predicted average adipocyte size. To examine this conventional model, we performed data augment described in Section 3.2 so that 891 samples were used to calculate the parameter slope and obtain the best fitting using the polynomial regression model. The parameters a, b, and c from the best fitting of the polynomial regression model was tested on the rest 33 samples (Section 3.2). By comparing the predicted adipocyte size with the result from the gold standard histology, the error for each sample in the test set was computed. With the results from all the samples in the test set, the MAE and MRE were quantified for the conventional method of PASA, and compared to the outcomes from deep learning. The errors of using the PASA method and deep learning method combining all frequency components were listed in Table 4. As we can see, deep learning method provided better accuracy compared to conventional PASA method.

Table 4. Comparison between deep learning method and photoacoustic spectral analysis (PASA) method.

Method	Deep Learning	PASA
MAE/μm^2	394.19	891.20
MRE	9.30%	22.14%

4. Conclusions

In this research, we explored the feasibility of characterizing adipocyte size in human adipose tissue using PA measurement combined with a deep learning method. In the experiments on *ex vivo* human adipose tissues, we first studied different networks with various numbers of layers and the results indicated that the network depth has a great influence on performance of network. As shown in Figure 4, the 7-layer network yielded the best performance and resulted in a nearly 10% decrease of MRE compared with the 2-layer network when training on the 12.5 to 16.5 MHz spectral band. When trained on different spectral bands and the entire spectral band, our network found out the optimally fitted relationship between the PA signal spectrum and the adipocyte size measured by histology. The most sensitive spectral band turned out to be relative to adipocyte size which can be evaluated by Equation (5). PA signals are characterized with a broad-band property; combining all spectral components resulted in an MRE decrease of 3.48% compared with training on the most sensitive spectral band in our study. Compared to conventional PASA method, using a deep learning method to fit the nonlinear relationship can better utilize the rich information in the power spectrum and an improvement of 12.84 % on MRE was achieved. The results of our research show the validity of analyzing PA signals in the frequency domain using deep learning which can be a novel method for quantitative and noninvasive evaluation of biological tissues, e.g., characterizing human adipose tissue cellular phenotype in the context of the clinical management of obese patients. In our further study, a larger data set will be built and a method of combining multiple wavelengths, instead of just using 1210 nm laser illumination, will be tested to improve accuracy of evaluation for clinical application.

Author Contributions: Conceptualization and methodology, J.Y. and X.W.; Data Curation, M.C., T.F., A.R.W., N.A.B., C.N.L. and R.W.O.; Software, X.M. and C.M.; Formal Analysis, X.M.; Writing, X.M. and C.M.; Supervision, Q.S., J.Y., Q.C. and X.W.

Funding: This research was funded by the National Key Research and Development Program of China (No. 2017YFC0111402), the Natural Science Funds of Jiangsu Province of China (No. BK20181256), and by NIH grants R01DK115190 (RWO, CNL) and R01DK097449 (RWO).

Appl. Sci. **2018**, *8*, 2178

Conflicts of Interest: The authors declare no conflict of interest.

References

1. Fleqal, K.M.; Carroll, M.D.; Oqden, C.L.; Curtin, L.R. Prevalence and trends in obesity among US adults, 1999–2008. *J. Am. Med. Assoc.* **2010**, *303*, 275–276. [CrossRef] [PubMed]
2. Nguyen, N.T.; Nguyen, X.-M.T.; Lane, J.; Wang, P. Relationship between obesity and diabetes in a us adult population: Findings from the national health and nutrition examination survey, 1999–2006. *Obes. Surg.* **2011**, *21*, 351–355. [CrossRef] [PubMed]
3. O'Connell, J.; Lynch, L.; Cawood, T.J.; Kwasnik, A.; Nolan, N.; Geoghegan, J.; McCormick, A.; O'Farrelly, C.; O'Shea, D. The relationship of omental and subcutaneous adipocyte size to metabolic disease in severe obesity. *PLoS ONE* **2010**, *5*, e9997. [CrossRef] [PubMed]
4. Skurk, T.; Alberti-Huber, C.; Herder, C.; Hauner, H. Relationship between adipocyte size and adipokine expression and secretion. *J. Clin. Endocrinol. Metab.* **2007**, *92*, 1023–1033. [CrossRef] [PubMed]
5. Muir, L.A.; Neeley, C.K.; Meyer, K.A.; Baker, N.A.; Brosius, A.M.; Washabaugh, A.R.; Varban, O.A.; Finks, J.F.; Zamarron, B.F.; Flesher, C.G.; et al. Adipose tissue fibrosis, hypertrophy, and hyperplasia: Correlations with diabetes in human obesity. *Obesity* **2016**, *24*, 597–605. [CrossRef] [PubMed]
6. Wang, X.; Pang, Y.; Ku, G.; Xie, X.; Stoica, G.; Wang, L.V. Noninvasive laser-induced photoacoustic tomography for structural and functional in vivo imaging of the brain. *Nat. Biotechnol.* **2003**, *21*, 803. [CrossRef] [PubMed]
7. Xu, M.; Wang, L.V. Photoacoustic imaging in biomedicine. *Rev. Sci. Instrum.* **2006**, *77*, 041101. [CrossRef]
8. Beard, P. Biomedical photoacoustic imaging. *Interface Focus* **2011**, *1*, 602–631. [CrossRef] [PubMed]
9. Kumon, R.E.; Deng, C.X.; Wang, X. Frequency-Domain analysis of photoacoustic imaging data from prostate adenocarcinoma tumors in a murine model. *Ultrasound Med. Biol.* **2011**, *37*, 834–839. [CrossRef] [PubMed]
10. Patterson, M.P.; Riley, C.B.; Kolios, M.C.; Whelan, W.M. Optoacoustic signal amplitude and frequency spectrum analysis laser heated bovine liver ex vivo. In Proceedings of the 2011 IEEE International Ultrasonics Symposium, Orlando, FL, USA, 18–21 October 2011; pp. 300–303.
11. Saha, R.K.; Kolios, M.C. A simulation study on photoacoustic signals from red blood cells. *J. Acoust. Soc. Am.* **2011**, *129*, 2935–2943. [CrossRef] [PubMed]
12. Xu, G.; Dar, I.A.; Tao, C.; Liu, X.; Deng, C.X.; Wang, X. Photoacoustic spectrum analysis for microstructure characterization in biological tissue: A feasibility study. *Appl. Phys. Lett.* **2012**, *101*, 221102. [CrossRef] [PubMed]
13. Xu, G.; Meng, Z.-X.; Lin, J.D.; Yuan, J.; Carson, P.L.; Joshi, B.; Wang, X. The functional pitch of an organ: Quantification of tissue texture with photoacoustic spectrum analysis. *Radiology* **2014**, *271*, 248–254. [CrossRef] [PubMed]
14. Feng, T.; Perosky, J.E.; Kozloff, K.M.; Xu, G.; Cheng, Q.; Du, S.; Yuan, J.; Deng, C.X.; Wang, X. Characterization of bone microstructure using photoacoustic spectrum analysis. *Opt. Express* **2015**, *23*, 25217–25224. [CrossRef] [PubMed]
15. Xu, G.; Meng, Z.-X.; Lin, J.-D.; Deng, C.X.; Carson, P.L.; Fowlkes, J.B.; Tao, C.; Liu, X.; Wang, X. High resolution physio-chemical tissue analysis: Towards noninvasive in vivo biopsy. *Sci. Rep.* **2016**, *6*, 16937. [CrossRef] [PubMed]
16. Feng, T.; Li, Q.; Zhang, C.; Xu, G.; Guo, L.J.; Yuan, J.; Wang, X. Characterizing cellular morphology by photoacoustic spectrum analysis with an ultra-broadband optical ultrasonic detector. *Opt. Express* **2016**, *24*, 19853–19862. [CrossRef] [PubMed]
17. Cao, M.; Zhu, Y.; O'Rourke, R.; Wang, H.; Yuan, J.; Cheng, Q.; Xu, G.; Wang, X.; Carson, P. Adipocyte property evaluation with photoacoustic spectrum analysis: A feasibility study on human tissues. In Proceedings of the SPIE BiOS, San Francisco, CA, USA, 3 March 2017; p. 6.
18. Lizzi, F.L.; Greenebaum, M.; Feleppa, E.J.; Elbaum, M.; Coleman, D.J. Theoretical framework for spectrum analysis in ultrasonic tissue characterization. *J. Acoust. Soc. Am.* **1983**, *73*, 1366–1373. [CrossRef] [PubMed]
19. Lizzi, F.L.; Feleppa, E.J.; Kaisar Alam, S.; Deng, C.X. Ultrasonic spectrum analysis for tissue evaluation. *Pattern Recognit. Lett.* **2003**, *24*, 637–658. [CrossRef]
20. Dreyfus, S.E. Artificial neural networks, back propagation, and the Kelley-Bryson gradient procedure. *J. Guid. Control Dyn.* **1990**, *13*, 926–928. [CrossRef]

21. Bengio, Y.; Courville, A.; Vincent, P. Representation learning: A review and new perspectives. *IEEE Trans. Pattern Anal. Mach. Intell.* **2013**, *35*, 1798–1828. [CrossRef] [PubMed]
22. Alipanahi, B.; Delong, A.; Weirauch, M.T.; Frey, B.J. Predicting the sequence specificities of DNA- and RNA-binding proteins by deep learning. *Nat. Biotechnol.* **2015**, *33*, 831. [CrossRef] [PubMed]
23. Plis, S.M.; Hjelm, D.R.; Salakhutdinov, R.; Allen, E.A.; Bockholt, H.J.; Long, J.D.; Johnson, H.J.; Paulsen, J.S.; Turner, J.A.; Calhoun, V.D. Deep learning for neuroimaging: A validation study. *Front. Neurosci.* **2014**, *8*, 229. [CrossRef] [PubMed]
24. Gulshan, V.; Peng, L.; Coram, M.; Stumpe, M.C.; Wu, D.; Narayanaswamy, A.; Venugopalan, S.; Widner, K.; Madams, T.; Cuadros, J.; et al. Development and Validation of a Deep Learning Algorithm for Detection of Diabetic Retinopathy in Retinal Fundus Photographs. *J. Am. Med. Assoc.* **2016**, *316*, 2402–2410. [CrossRef] [PubMed]
25. Tsai, C.-L.; Chen, J.-C.; Wang, W.-J. Near-infrared absorption property of biological soft tissue constituents. *J. Med. Biol. Eng.* **2001**, *21*, 7–14.
26. Perez, L.; Wang, J. The effectiveness of data augmentation in image classification using deep learning. *arXiv* **2017**; arXiv:1712.04621.
27. Takahashi, N.; Gygli, M.; Pfister, B.; Van Gool, L. Deep convolutional neural networks and data augmentation for acoustic event detection. In Proceedings of the Interspeech 2017, San Francisco, CA, USA, 8–12 September 2016.
28. Hammer, B.E. Physical Properties of Tissues. *Mosc. Univ. Math. Bull.* **1991**, *54*, 73. [CrossRef]
29. Barham, R.H.; Drane, W. An Algorithm for Least Squares Estimation of Nonlinear Parameters When Some of the Parameters Are Linear. *Technometrics* **1972**, *14*, 757–766. [CrossRef]
30. Hawkins, D.M. The Problem of Overfitting. *J. Chem. Inf. Comput. Sci.* **2004**, *44*, 1–12. [CrossRef] [PubMed]

Article

Biomedical Photoacoustic Imaging Optimization with Deconvolution and EMD Reconstruction

Chengwen Guo [1], Yingna Chen [2], Jie Yuan [1,*], Yunhao Zhu [1], Qian Cheng [2] and Xueding Wang [2]

[1] School of Electronic Science and Engineering, Nanjing University, Nanjing 210093, China;
    guochevonne@163.com (C.G.); zyh6557@126.com (Y.Z.)
[2] Institute of Acoustic, Tongji University, Shanghai 200092, China; 1610532@tongji.edu.cn (Y.C.);
    q.cheng@tongji.edu.cn (Q.C.); xdwang@umich.edu (X.W.)
*   Correspondence: yuanjie@nju.edu.cn; Tel.: +86-186-2518-6370

Received: 12 October 2018; Accepted: 21 October 2018; Published: 1 November 2018

Abstract: A photoacoustic (PA) signal of an ideal optical absorbing particle is a single N-shape wave. PA signals are a combination of several individual N-shape waves. However, the N-shape wave basis leads to aliasing between adjacent micro-structures, which deteriorates the quality of final PA images. In this paper, we propose an image optimization method by processing raw PA signals with deconvolution and empirical mode decomposition (EMD). During the deconvolution procedure, the raw PA signals are de-convolved with a system dependent deconvolution kernel, which is measured in advance. EMD is subsequently adopted to further process the PA signals adaptively with two restrictive conditions: positive polarity and spectrum consistency. With this method, signal aliasing is alleviated, and the micro-structures and detail information, previously buried in the reconstructing images, can now be revealed. To validate our proposed method, numerical simulations and phantom studies are implemented, and reconstructed images are used for illustration.

Keywords: photoacoustic imaging; signal processing; deconvolution; empirical mode decomposition; signal deconvolution

---

1. Introduction

Photoacoustic (PA) imaging, an emerging biomedical imaging modality based on PA effect, has been developed extensively in recent years. During PA imaging, laser pulses energy are delivered into biological tissues, leading to transient thermoelastic expansion and thus wideband ultrasound emission [1]. The generated ultrasound waves propagate to the surface where detected by ultrasound transducers [2], and are reconstructed into an image. As PA images can reveal pathology features and physiological structures according to the specific optical absorption distribution of biological tissues, the diagnoses on the tissue differs in physiological properties a lot, such as breast cancer diagnosis [3], and hemodynamics monitoring [4,5], making PA imaging a promising and high potential imaging modality.

Since typical detected PA signals of ideal optical absorbing particle is a single bipolar N-shape pulse [6,7], PA signals of a complicated biological tissue can be considered to be the combination of individual N-shape pulses. However, the N-shape wave basis results in two problems: the first problem is aliasing between adjacent micro-structures. The signal of a tiny target can be affected and even buried by the bipolar signal of a large target at a short distance, leading to unexpected aliasing and distortion in the final image. The second problem is that the existence of the N-shape wave complicates subsequent imaging work. When reconstructing images, the envelope of signal must transform the negative part into positive, as the negative part possesses significant information as well. Both drawbacks could deteriorate the quality of PA images.

Therefore, the processing on an N-shape wave can be significant in alleviating the drawbacks and has been investigated a number of studies. Li [7], identified some properties about the N-shape wave and introduced processing methods, including wavelet and deconvolution. Ermilov [8,9], transformed the bipolar N-shape pressure pulse to the monopolar pulse using wavelet transform in order to get the signal suitable for tomographic reconstruction. The deconvolution method can convert the bipolar N-shape wave to the monopolar wave and is rarely applied to processing signals, and thereby raising our interest. Traditionally, the deconvolution method is often applied to image reconstruction and image processing. For instance, Kruger [10] and Gamelin [11] used deconvolution methods in PA image reconstruction. Cai [12] improved the image resolution of PA microscopy by using deconvolution method on images. Recently, Nagaoka [13] proposed a reconstruction method to improve axial resolution through the suppression of the time side lobes in PA tomography by Wiener filtering.

In this paper, we propose using the image optimization method to process raw PA signals using signal deconvolution and empirical mode decomposition (EMD). This method mainly relies on signal deconvolution. However, as unexpected artifacts appeared around the imaging target only with signal deconvolution, EMD was instead adopted into our method as a subsequent step to further improve the reconstructed image quality. Moreover, we averaged multi-sampling raw PA signals with time shift correction by cross-correlation method as a preprocessing step in order to raise the signal-to-noise ratio (SNR) and enhance the performance of signal deconvolution.

2. Methods

Given that the detected PA signals have low SNR that deteriorate the quality of PA images, de-noising operation is quite common as a pre-processing step. Theoretically, averaging signals of multiple frames can eliminate random background noise and raise the SNR. However, non-negligible time shifts among signals of different frames always exist for a variety of reasons, including the slight fluctuation of the water around the imaging target, the small movement from a living biological imaging target, and system defect (possibly random delay generated by inner circuit). To solve this issue, we adopted an effective averaging method to adjust time shifts based on cross-correlation method [14,15].

First, we calculated the time shifts $(\Delta m_1, \Delta m_2, \cdots)$ between the reference frame and other frames by cross-correlation. For discrete PA signals $p_1(n)$ and $p_2(n)$, an estimate of the cross-correlation function is defined as [16]:

$$\hat{R}_{p_1 p_2}(m) = \begin{cases} \frac{1}{L-|m|} \sum_{n=0}^{L-m-1} p_1(n+m) p_2^*(n), & m \geq 0 \\ \hat{R}_{p_1 p_2}(-m), & m < 0 \end{cases}, \qquad (1)$$

where $p_2^*(n)$ is the complex conjugate of $p_2(n)$, L is the maximum length of $p_1(n)$ and $p_2(n)$, the normalization coefficient $1/(L-|m|)$ can avoid the zero bias inherent in the standard cross-correlation function. According to this equation, the value of time shift Δm should equal to the value of m when the cross-correlation function $\hat{R}_{p_1 p_2}(m)$ is of maximum value.

Assuming $p_1(n)$ is the uniform reference, $p_2(n)$, the signals of all the rest frames need to be shifted to the length of their corresponding $|\Delta m|$ forwards or backwards based on the sign of Δm. After that, shifted signals of multiple frames that all align is acquired, so we can obtain the final de-noised signal $\hat{p}(n)$ by simply averaging them all.

Mathematically, the convolution result of a single wave and positive pulses with different delay is the superposition of many same-shaped waves with different delay, and the specific size of each wave depends on the size of the corresponding positive pulse. Hence, we assumed that PA signal is the convolution result of a series of specific monopolar pulses and a single N-shape wave. With that assumption, the monopolar signal, which contains the amplitude information and time information, can theoretically be acquired through deconvolution. As we mentioned in the former section, the N-shape wave basis has defects that will reduce the quality of PA images, while the de-convolved monopolar signals possess equally important information for imaging as raw PA signals and perform better than raw N-shape PA signals when reconstructed into images. We intended to acquire de-convolved signals to replace raw PA signals.

Generally, the discrete detected PA signal s(n) can be represented in convolution form as:

$$s(n) = o(n) * h(n) + n(n), \qquad (2)$$

where $*$ denotes convolution, $h(n)$ is the kernel of this signal convolution system, s(n) is the detected raw signal, $o(n)$ refers to the signal which we aim to acquire and $n(n)$ represents the additive noise. In our method, we took $h(n)$ as a normalized single N-shape pulse which is acquired by detecting an approximately ideal optical absorbing particle, specifically, the smallest source which the PA data acquisition system can measure. For example, assuming that velocity is 1540 m/s, an L7-4 ultrasound probe whose center frequency is 5.208 MHz can detect the minimum source is approximately a diameter of 0.296 mm, so in this case, the kernel $h(n)$ ought to be acquired by detecting the source whose size is closest to and bigger than 0.296 mm.

The deconvolution algorithm we used was the Wiener deconvolution [17]. A typical simulated signal was used for demonstration. As shown in Figure 1b, the signal was converted from a bipolar signal to monopolar signal after deconvolution. However, directly using de-convolved signal to reconstruct images creates defects. As shown in Figure 1c, when reconstructing images by conventional delay-and-sum algorithm, assuming the velocity is uniform, the value corresponding with the position of the imaging target was also added to the other positions of the same distance to the transducer element as imaging target. Normally, artifacts can be restrained as the signal has both positive part and negative part to cancel each other out. Since the de-convolved signal is monopolar, unexpected artifacts will appear around the imaging object because amplitude cannot be cancelled out when summing up as they are all positive. To alleviate this phenomenon and further optimize the images, EMD was adopted subsequent to signal deconvolution.

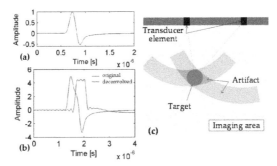

Figure 1. (**a**) Diagram of the normalized deconvolution kernel $h(n)$ in simulation; (**b**) comparison of simulated signal before and after deconvolution; (**c**) illustration of the widespread artifacts forming process in delay-and-sum algorithm.

The principle of EMD [18], is to decompose a signal into several intrinsic mode functions (IMFs) without setting any basis function in advance, and each of them reserves the local features of the original signal with different time scales. Given that different IMFs possess different features, we were able to rebuild new signals to meet our requirements by depressing some of the IMFs and enhancing other IMFs. Specifically, we applied weight coefficients to each IMF and refactor new signal automatically under the constraints which is set in advance. In our proposed method, we apply the de-convolved signal to EMD and refactor it with two constraints, positive polarity and spectrum consistence. Given that EMD can only process one-dimensional signals at a time, a one-dimensional signal $o_1(n)$ (detected by a single transducer element of the probe) of de-convolved signal o(n) should be considered as an example, the procedure [19] is described in detail as follows:

Step 1: Use EMD method to decompose signal $o_1(n)$ into k IMFs (k is a changeable number) and the residue:

$$o_1(n) = \sum_{i=1}^{k} IMF_i + res(n),\tag{3}$$

Step 2: Set weight coefficient value of each IMF to refactor a set of new signals:

$$new_{o_1(n)} = [a_1\ a_2\ \cdots\ a_k] \times \begin{bmatrix} IMF_1 \\ IMF_2 \\ \vdots \\ IMF_k \end{bmatrix},\tag{4}$$

where $a = [a_1\ a_2\ \cdots\ a_k]$ refers to the weight coefficients of IMFs. As the number of IMFs is k, the number of weight coefficient is also k and each weight coefficient corresponds with an IMF. The value of each weight coefficient ranges from $[0, 1]$ with step 0.1, so that the number of iteration is 11^k. The coefficient of residue remains 1.

Step 3: Calculate two constraints in each iteration:

The first constraint is positive polarity. As the original de-convolved signal is monopolar, positive polarity is set to avoid distortion between the refactor signal and the original signal. Positive polarity is defined as:

$$p = \frac{\min(new\_o_1(n))}{\max(new\_o_1(n))} > \text{threshold},\tag{5}$$

where the threshold here should equal the minimum amplitude of $o_1(n)$ divided by maximum amplitude of $o_1(n)$. Within each iteration, the second constraint can be calculated if the refactored signal satisfies the positive polarity. Otherwise, this iteration should be skipped and to begin the next iteration.

The second constraint is spectrum consistence. As the high frequency components usually retain detailed information, we applied the spectrum consistence to refactoring a new signal, with more detail and less aliasing. Spectrum consistence is defined as:

$$c = \sum_{\omega=f_1}^{f_2} f(\omega) \Big/ \sum_{\omega=f_0}^{f_1} f(\omega),\ \omega < \frac{f_s}{2},\tag{6}$$

where $f(\omega)$ is the FFT result of refactored signal, f_s refers to the sampling frequency. As band-limited ultrasound transducers can only receive the signal with a limited frequency band, f_0 refers to the low cut-off frequency, f_2 refers to the high cut-off frequency and f_1 is the middle frequency point of the frequency band. We assume that the high frequency component and the low frequency component is divided by f_1. Set the initial $c = 0$, and if the c of this iteration is bigger than the former one, the weight coefficient of this iteration should be adopted and the former c should be substituted by the new c.

Step 4: repeat the iterations in Step 2 and Step 3 until it traverses all the weight coefficients, and we can then assume that the $new\_o_1(n)$, which satisfies the positive polarity and obtains the biggest spectrum consistence c is the best refactored signal.

3. Results

3.1. Numerical Simulations

We implemented numerical simulations based on the MATLAB (The MathWorks, Natick, MA, USA) k-Wave toolbox [20] using a HP server (Hewlett-Packard, Palo Alto, CA, USA), which has 2 Intel (Intel Corporation, Santa Clara, CA, USA) Xeon (R) X5670 CPU working at 2.93 GHz and 72.0 GB RAM. PA data were detected by 128 linear ultrasound element transducers, which distributed on the upper side of the numeric phantom. The speed of sound was set to 1540 m/s. In the simulation, two numeric phantoms were used to validate the effectiveness of our proposed method. To prevent the interference of the reconstruction method in k-Wave toolbox, we used the basic delay-and-sum reconstruction algorithm to reconstruct each image in our study.

First, the deconvolution kernel $h(n)$ was acquired from a single transducer element on the upper side, which detects the numeric phantom that only consists of a point source of radius 1 in the middle position. The normalized $h(n)$, shown in Figure 1a performs as the uniform deconvolution kernel in our simulations.

To illustrate the deformation of signal and the direct results in reconstructed image clearly, we used a simple numeric phantom as seen in Figure 2a, which consists of a large point source in the middle and four small point sources around. Meanwhile, to make the simulation more realistic, we used a vessel-like image as seen in Figure 3a to simulate the complex numeric phantom.

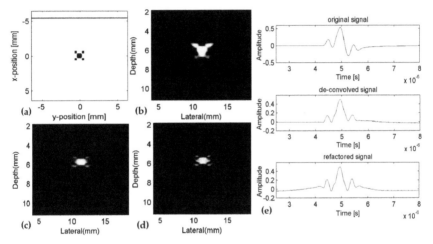

Figure 2. Results of the signal processing and reconstruction in simple numeric phantom. (**a**) the set position of source and sensors; (**b**) reconstructed image of original signal; (**c**) reconstructed image of de-convolved signal; (**d**) reconstructed image of final processed signal; (**e**) signal processing result.

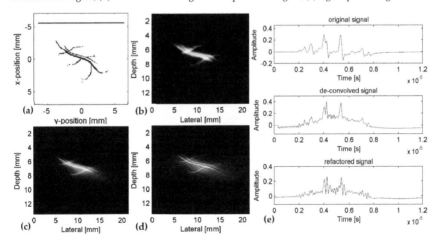

Figure 3. Results of the signal processing and reconstruction in complex numeric phantom. (**a**) the set position of source and sensors; (**b**) reconstructed image of original signal; (**c**) reconstructed image of de-convolved signal; (**d**) reconstructed image of final processed signal; (**e**) signal processing result.

As observed in Figures 2e and 3e, bipolar N-shape signals are converted to monopolar signals after processing and more detailed information appears, especially in Figure 3e. The small N-shape waves affected by the negative part of the big N-shape waves became apparent. From the reconstructed images in Figures 2 and 3, we can observe that the original signals result in aliasing and blurring in reconstruction (as Figures 2b and 3b), the five dots and the two main branches are hard to distinguish. Moreover, the existence of N-shape wave leads to distortion in the image. Using Figure 2b as an example, the two points in the lower side are obviously weaker than the two points in the upper side, and on the other hand, the position of the two points in the lower side slightly moves downward compared with the set position (Figure 2a). This is because the wave after big waves is affected and even buried by the negative part of the big waves. But by performing deconvolution on signals (Figures 2c and 3c), aliasing is alleviated, and distortion is corrected. After processing by EMD (Figures 2d and 3d), the edge sharpness and resolution of the images were further improved. With the gap widening between main imaging targets, the five dots in Figure 2 and the branches in Figure 3, are distinguishable and separated now.

Since the original PA signal is bipolar, using an envelope is necessary when reconstructing images by means of delay-and-sum algorithm. However, this step could be left out due to the positive polarity of the refactored signal, which further improves image quality and increases efficiency.

3.2. Experiments

The schematic of our experimental setup is shown in Figure 4a. A tunable optical parametric oscillator (OPO) system (Phocus Mobile, Opotek, CA, USA) is pumped by a Nd; YAG laser (Brilliant B, Bigsky, MT, USA) is employed as the laser source. The laser works at the wavelength of 720 nm with pulse repetition rate of 10 Hz and pulse duration of 5.5 ns. A Verasonics system (Verasonics Inc., Kirkland, WA, USA) is controlled by a computer to receive raw PA data. An ultrasound linear array (L7-4, Philips, Amsterdam, The Netherlands) with 128 elements, 5.208 MHz center frequency and 0.298 mm element pitch was used. Ultrasonic coupling agent is used as the acoustic transmission medium between phantom and ultrasound probe.

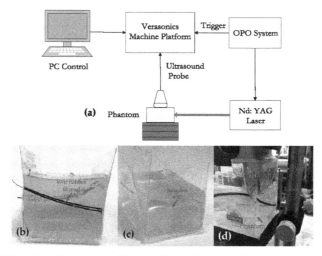

Figure 4. (**a**) Schematic of the experimental setup; (**b**) the phantom used as imaging target (two rubber threads); (**c**) the phantom to acquire the deconvolution kernel (one tungsten wire); (**d**) a photo of the experiment setup.

The phantoms used in this experiment were made of porcine gelatin with 8% concentration as shown in Figure 4b,c. Figure 4b is the phantom used as the imaging target in our experiment which embedded with two rubber threads of diameter 1 mm and Figure 4d shows the placement of phantom, probe and laser in this experiment. The arrows in Figure 4b,c both point to the position where a probe acquired a signal of each phantom while carrying out the experiment. During the process of signal acquisition, each set of signals contained approximately 200 frames for a continuous period to average.

The deconvolution kernel is acquired from the middle transducer element of the probe, which detected a single tungsten wire of diameter 0.3 mm (as Figure 4c); approximately the smallest size L7-4 probe can measure. The normalized deconvolution kernel (as shown in Figure 5), averaged from 200 frame signals, was used as the deconvolution kernel in this experiment.

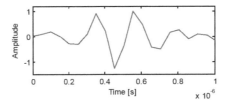

Figure 5. Diagram of the normalized deconvolution kernel in the experiment.

The comparison of the reconstructed result is shown in Figure 6. We can observe that the two oblique rubber wires are too close to look separated in the image reconstructed with original PA signal, as shown in Figure 6a. Without knowing the imaging targets in advance, it is difficult to distinguish the structure visually. After applying our proposed method, as shown in Figure 6b, the two lines are completely separated, and the realistic inner structures of this phantom can be more clearly identified. Meanwhile, image resolution is enhanced, and the artifact is slightly reduced.

To quantify the width of gaps in Figure 6 and compared to the realistic width of gap in Figure 4b, we measured the width of gaps in Figure 6. In Figure 4b, the diameter of the two rubber wires in the phantom are both 1 mm, we can observe that the realistic gap, where the arrow mark is pointing to, is approximately 0.4 mm. In respect of the reconstructed image of the original signal as shown in Figure 6a, the width of the gap is about 0.08 mm. For the reconstructed image of the processed signal as shown in Figure 6b, the width of the gap is about 0.3 mm. From the quantification, we can find that the image of processed signal more aptly indicates the accurate width and position of realistic targets.

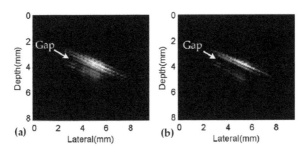

Figure 6. Reconstruction results of the experiment. (**a**) the reconstructed image of original signal; (**b**) the reconstructed image of processed signal.

4. Discussion and Conclusions

From the contrast of signals in the simulation before and after processing, the bipolar signals are converted into non-negative signals, whereas the processed signals in the experiment are relatively monopolar, but not strictly monopolar. The deconvolution result is sensitive to noise, which is inevitable in the signal acquisition procedure. Meanwhile, the relatively monopolar signals have some benefit in reducing artifacts in the final reconstructed image. Widespread artifacts as shown in

Figure 2d appear around imaging targets when using the strictly monopolar signal, whereas artifacts can be slightly alleviated if the relatively monopolar signal are used.

In this work, we demonstrate an image optimization method by processing the raw PA signals using signal deconvolution and EMD. Using this method, raw bipolar signals are converted into quasi-monopolar signals and more detailed information can be obtained to alleviate the signal aliasing phenomenon. With the processed signals, PA image can be optimized, especially in resolution and edge sharpness. Moreover, the micro-structures, which were previously hidden and buried by large imaging targets, can now be revealed and separated. Both numeric simulations and phantom study illustrate the effectiveness of this method. As this method can help to distinguish between close objects and reveal detail information, it can potentially help with diagnoses, especially on tissues with micro-structures.

Author Contributions: Conceptualization, C.G. and J.Y.; data curation, C.G.; formal analysis, C.G.; funding acquisition, J.Y.; investigation, C.G.; methodology, C.G., J.Y. and Y.Z.; project administration, J.Y.; resources, J.Y., Q.C. and X.W.; software, C.G.; supervision, J.Y.; validation, C.G., Y.C. and Q.C.; visualization, C.G.; writing–original draft, C.G.; writing–review and editing, J.Y. and Y.Z.

Funding: This research was funded by the National Key Research and Development Program of China (No. 2017YFC0111402) and the Natural Science Funds of Jiangsu Province of China (No. BK20181256).

Conflicts of Interest: The authors declare no conflicts of interest.

References

1. Xu, M.; Wang, L.V. Photoacoustic imaging in biomedicine. *Rev. Sci. Instrum.* **2006**, *77*, 041101. [CrossRef]
2. Beard, P. Biomedical photoacoustic imaging. *Interface Focus* **2011**, *1*, 602–631. [CrossRef] [PubMed]
3. Ku, G.; Fornage, B.D.; Jin, X.; Xu, M.; Hunt, K.K.; Wang, L.V. Thermoacoustic and photoacoustic tomography of thick biological tissues toward breast imaging. *Technol. Cancer Res. Treat.* **2005**, *4*, 559–565. [CrossRef] [PubMed]
4. Wang, X.; Pang, Y.; Ku, G.; Xie, X.; Stoica, G.; Wang, L.V. Noninvasive laser-induced photoacoustic tomography for structural and functional in vivo imaging of the brain. *Nat. Biotechnol.* **2003**, *21*, 803–806. [CrossRef] [PubMed]
5. Wang, X.; Xie, X.; Ku, G.; Wang, L.V.; Stoica, G. Noninvasive imaging of hemoglobin concentration and oxygenation in the rat brain using high-resolution photoacoustic tomography. *J. Biomed. Opt.* **2006**, *11*, 024015. [CrossRef] [PubMed]
6. Oraevsky, A.A.; Andreev, V.A.; Karabutov, A.A.; Esenaliev, R.O. Two-dimensional optoacoustic tomography: Transducer array and image reconstruction algorithm. In Proceedings of the BiOS '99 International Biomedical Optics Symposium, San Jose, CA, USA, 14 June 1999; p. 12.
7. Changhui, L.; Lihong, V.W. Photoacoustic tomography and sensing in biomedicine. *Phys. Med. Biol.* **2009**, *54*, R59.
8. Ermilov, S.A.; Gharieb, R.; Conjusteau, A.; Miller, T.; Mehta, K.; Oraevsky, A.A. Data processing and quasi-3D optoacoustic imaging of tumors in the breast using a linear arc-shaped array of ultrasonic transducers. In Proceedings of the SPIE BiOS, San Jose, CA, USA, 28 February 2008; p. 10.
9. Ermilov, S.A.; Khamapirad, T.; Conjusteau, A.; Leonard, M.H.; Lacewell, R.; Mehta, K.; Miller, T.; Oraevsky, A.A. Laser optoacoustic imaging system for detection of breast cancer. *J. Biomed. Opt.* **2009**, *14*, 024007. [CrossRef] [PubMed]
10. Kruger, R.A.; Kiser, W.L.; Reinecke, D.R.; Kruger, G.A. Thermoacoustic computed tomography using a conventional linear transducer array. *Med. Phys.* **2003**, *30*, 856–860. [CrossRef] [PubMed]
11. Gamelin, J.; Maurudis, A.; Aguirre, A.; Huang, F.; Guo, P.; Wang, L.V.; Zhu, Q. A real-time photoacoustic tomography system for small animals. *Opt. Express* **2009**, *17*, 10489–10498. [CrossRef] [PubMed]
12. Cai, D.; Li, Z.; Chen, S.-L. In vivo deconvolution acoustic-resolution photoacoustic microscopy in three dimensions. *Biomed. Opt. Express* **2016**, *7*, 369–380. [CrossRef] [PubMed]
13. Ryo, N.; Shin, Y.; Shin-ichiro, U.; Yoshifumi, S. Basic study of improvement of axial resolution and suppression of time side lobe by phase-corrected wiener filtering in photoacoustic tomography. *Jpn. J. Appl. Phys.* **2018**, *57*, 07LD11. [CrossRef]

14. Li, W.; Zhang, L.; Yuan, J.; Liu, X.; Xu, G.; Wang, X.; Carson, P.L. Novel image optimization on photoacoustic tomography. In Proceedings of the 2014 IEEE International Ultrasonics Symposium, Chicago, IL, USA, 3–6 September 2014; pp. 1280–1283.
15. Cao, M.; Feng, T.; Yuan, J.; Xu, G.; Wang, X.; Carson, P.L. Spread spectrum photoacoustic tomography with image optimization. *IEEE Trans. Biomed. Circuits Syst.* **2017**, *11*, 411–419. [CrossRef] [PubMed]
16. Brunker, J.; Beard, P. Pulsed photoacoustic doppler flowmetry using time-domain cross-correlation: Accuracy, resolution and scalability. *J. Acoust. Soc. Am.* **2012**, *132*, 1780–1791. [CrossRef] [PubMed]
17. Gonzales, R.C.; Woods, R.E. *Digital Image Processing*; Addison & Wesley Publishing Company: Reading, MA, USA, 1992.
18. Huang, N.E.; Shen, Z.; Long, S.R.; Wu, M.C.; Shih, H.H.; Zheng, Q.; Yen, N.-C.; Tung, C.C.; Liu, H.H. The empirical mode decomposition and the hilbert spectrum for nonlinear and non-stationary time series analysis. *Proc. R. Soc. Lond. Ser. A Math. Phys. Eng. Sci.* **1998**, *454*, 903. [CrossRef]
19. Zhu, Y.-H.; Yuan, J.; Pinter, S.Z.; Oliver, D.K.; Cheng, Q.; Wang, X.-D.; Tao, C.; Liu, X.-J.; Xu, G.; Carson, P.L. Adaptive optimization on ultrasonic transmission tomography-based temperature image for biomedical treatment. *Chin. Phys. B* **2017**, *26*, 064301. [CrossRef]
20. Treeby, B.E.; Cox, B.T. K-wave: Matlab toolbox for the simulation and reconstruction of photoacoustic wave fields. *J. Biomed. Opt.* **2010**, *15*, 021314. [CrossRef] [PubMed]

Article

A Novel Dictionary-Based Image Reconstruction for Photoacoustic Computed Tomography

Parsa Omidi [1,2], Mohsin Zafar [1], Moein Mozaffarzadeh [3], Ali Hariri [1,4], Xiangzhi Haung [1], Mahdi Orooji [3] and Mohammadreza Nasiriavanaki [1,*]

1   Department of Biomedical Engineering, Wayne State University, Detroit, MI 48202, USA;
     mrn.avanaki@wayne.edu (P.O.); mohsin.zafar@wayne.edu (M.Z.); haririali92@gmail.com (A.H.);
     xianhuan@umich.edu (X.H.)
2   Department of Biomedical Engineering, University of Western Ontario, London, ON N6A 3K7, Canada
3   Department of Biomedical Engineering, Tarbiat Modares University, Tehran 14115-111, Iran;
     moein.mfh@gmail.com (M.M.); morooji@gmail.com (M.O.)
4   Department of NanoEngineering, University of California, San Diego, 9500 Gilman Drive, La Jolla,
     CA 92092, USA
*   Correspondence: ft5257@wayne.edu; Tel.: +1-313-577-0703

Received: 9 July 2018 ; Accepted: 4 September 2018 ; Published: 6 September 2018

Abstract: One of the major concerns in photoacoustic computed tomography (PACT) is obtaining a high-quality image using the minimum number of ultrasound transducers/view angles. This issue is of importance when a cost-effective PACT system is needed. On the other hand, analytical reconstruction algorithms such as back projection (BP) and time reversal, when a limited number of view angles is used, cause artifacts in the reconstructed image. Iterative algorithms provide a higher image quality, compared to BP, due to a model used for image reconstruction. The performance of the model can be further improved using the sparsity concept. In this paper, we propose using a novel sparse dictionary to capture important features of the photoacoustic signal and eliminate the artifacts while few transducers is used. Our dictionary is an optimum combination of Wavelet Transform (WT), Discrete Cosine Transform (DCT), and Total Variation (TV). We utilize two quality assessment metrics including peak signal-to-noise ratio and edge preservation index to quantitatively evaluate the reconstructed images. The results show that the proposed method can generate high-quality images having fewer artifacts and preserved edges, when fewer view angles are used for reconstruction in PACT.

Keywords: photoacoustic imaging; image quality assessment; image formation theory; image reconstruction techniques; sparsity

1. Introduction

In a photoacoustic computed tomography (PACT) configuration, the ultrasonic waves are collected using ultrasonic transducers placed all around the tissue. The waves are then processed through a reconstruction algorithm, and an image is generated [1–4]. The number of transducers/view angles in PACT is directly proportional to the quality of the reconstructed image. In the past several years, researchers in the field of PACT have focused on two main topics: the design of the imaging system and image reconstruction algorithms [5–9]. In the case of system design, most of the attention has been on the arrangement of the transducers. The simplest proposed configuration has been using a single transducer swiping around the targeted object, and a more complicated one is to use a transducer array that covers either 360 degrees around the object (ring shape) or some parts of it. The type of transducers used in the configuration of PACT can be spherical, cylindrical, or flat [10–13]. The expansion of this concept is beyond the scope of this study. The literature cited here is a limited amount of PACT work.

For a more complete list, please read [14]. Readers are referred to Figure 1 of [15] and Figure 11 of [16] for further explanation of PACT imaging systems.

Several reconstruction algorithms have been designed for PACT, where a circular or linear array has been used for data acquisition [17–19]. Back-Projection (BP) and its derivations e.g., filtered back projection (FBP), could be considered as the most eminent PACT reconstruction algorithm due to their simple implementation [20]. They use the fact that pressures propagating from an acoustic source reach the detectors at different time delays, which depends on the speed of sound, as well as the distance between the source and the detectors [6,8,21].

The requirement in the BP algorithm is to input a large number of signals collected from different view angles. As mentioned, for a circular detection, the signals can be collected either using a single transducer rotating around the object or a ring shape transducer array. Such systems are either too expensive (due to the transducer ring array) or have a poor temporal resolution (due to using single transducer). To address theses problems, iterative image reconstruction algorithms (known as model-based algorithms too) can be used where there is a model used for the reconstruction [22]. In other words, iterative algorithms build up a model to describe the relationship between the detected photoacoustic (PA) signals and the optical absorption distribution and iteratively reduce the artifacts [23–25].

One of the concepts which can be combined with the model-based algorithms is compressed sensing (CS) and sparsity [26]. Over the past few years, it has been shown that using sparsity with model-based image reconstruction algorithms can mitigate the artifacts caused by the limited view angles and improve the quality of the reconstructed PA images [24,27–29]. Donoho et al. proposed the CS theory in 2006, for the first time, which is based on a prior knowledge of unknowns [30]. This method can be used in data acquisition procedure of PACT [31], which leads to information reconstruction based on the convex optimization from some observations that seem highly incomplete [28]. This concept has been used in the reconstruction of magnetic resonance imaging (MRI) and Computed Tomography (CT)-MRI in order to reduce the scan time and have an inherent registration in space and time [32]. In addition, CS has been used for thermoacoustic imaging (TAI) [33].

Provost and Lesage [28] have used CS for PACT image reconstruction, where due to the advantages of sparse characteristic of a sample, fewer projections were used to reconstruct the optical absorption distribution map of the tissue [34]. Signal sparsification can be done in two different ways: (i) applying a dictionary learning and (ii) using standard transform functions. Even though dictionary learning-based methods outperform the standard transform functions, in order to make the signal sparser, they impose a higher complexity for reconstruction [35,36]. On the other hand, standard transform functions can simply translate the signal to a sparse shape. The Wavelet Transform (WT), Discrete Cosine Transform (DCT), and Total Variation (TV) are some of the popular sparsifying transforms. Such transformations result different sparse representations of the original signal. Similar to compression algorithms, the combination of the original signal in different transformation domains, can provide an image with an improved quality.

In this paper, we introduce a novel dictionary for sparse representation of PA signals in order to improve the quality of the reconstructed PA image. The proposed method consists of three sparsifying transformation functions: WT, DCT, and TV. Using the proposed method, users are able to highlight information provided by each of the functions. We compared the performance of the proposed algorithm with the performance of BP and sparse reconstruction based on different methods when the same number of view angles is used. We use two established quality assessment metrics—peak signal-to-noise ratio (PSNR) and edge preservation index (EPI)—for quantitative evaluation of the results. Quantitative and qualitative results indicate that the proposed method can be a proper option when we face a limited number of angle views.

2. Methods

2.1. Analytical Reconstruction

In a PACT system, after laser excitation, acoustic waves are generated. The waves are collected by ultrasonic transducers all around the imaging target, stored in a computer by a data acquisition (DAQ) card, and processed in a reconstruction algorithm. The acoustic wave generation is based on thermoelastic expansion effect. Based on the principle of acoustic theories, in a homogeneous medium, the pressure at position r and time t, $p(r, t)$, follows the thermoacoustic equation shown in Equation (1). In this equation, the initial pressure, $p_0(r)$, is generated by a short pulse, which can be mathematically considered as a delta function ($\delta(t)$) [20].

$$\nabla^2 p(r,t) - \frac{1}{c^2}\frac{\partial^2}{\partial t^2}p(r,t) = -p_0(r)\frac{d\delta(t)}{dt} \tag{1}$$

where c indicates the speed of sound. The pressure at (r, t), generated from the initial pressure at r', is given in Equation (2), known as the Forward Problem.

$$p(r,t) = \frac{\partial}{\partial t}\left[\frac{1}{4\pi c^3 t}\int dr' p_0(r')\delta(t - \frac{|r-r'|}{c})\right]. \tag{2}$$

In PACT, we look for initial pressure ($p_0(r')$) calculation using the pressure measured at different view angles/times ($p(r, t)$). To this end, Equation (3), known as the Backward Problem, would be used [27].

$$p_0(r) = \int \left[2p(r_0, \bar{t}) - 2\bar{t}\frac{\partial}{\partial \bar{t}}p(r_0, \bar{t})\right]\frac{d\Omega_0}{\Omega_0} \tag{3}$$

where $\bar{t} = |r - r_0|$, and $\frac{d\Omega_0}{\Omega_0}$ denotes the weight that must be allocated to the detected pressures by transducers [27]. A favorite reconstruction algorithm for the PACT system could be the one that generates a high-quality image with fewer number of view angles. Analytical algorithms such as BP and TV-based have an inherent limitation (necessity of a large number of view angles around the target object) for accurate estimation of the optical absorption. In other words, such algorithms cause artifacts in the reconstructed images when a limited number of view angles is available in the PACT system. One solution to address this problem is a model-based algorithm in which artifacts and noise are iteratively degraded using a model. It should be noted that reconstruction algorithms are mostly simplified, and the effects of transducer size, imaging noise, and directivity of sensors are not considered in the reconstruction procedure. All these assumptions lead to negative effects called artifacts in this paper.

2.2. The Proposed Method

We model the procedure of PACT using $b = Ax$, where A is the measurement matrix, and x is the assumed phantom in the forward problem and the reconstructed image in the backward problem. b is the detected signals by the ultrasound transducers. When the number of unknowns (number of pixels of the PA image) is greater than the number of equations (number of recorded samples), which usually happens in PACT, especially when a low number of view angles is used for decreasing the data acquisition time, we have an underdetermined system. In this case, there is no exact solution.

One commonly used method to solve an underdetermined problem is the least square technique, giving $x_{est} = (A^T A)^{(-1)} A^T b$. The least square method uses the error minimization to obtain a solution, but the answer is not the best and the most accurate one that can be obtained. The sparse component analysis and sparse representation of signals can be used to solve the underdetermined problem of the PA image reconstruction. Having a prior knowledge, we can promote sparsity using l_0, l_1, and l_p ($0 < p < 1$) norms to obtain a more accurate solution [30]. In other words, we can improve the image quality by assuming that the imaging target is sparse.

Targets in PA imaging are composed of many singularities. A singularity is the point of an exceptional set where it fails to be well-behaved in differentiability. These singularities can be considered as non-zero values of x. Such an assumption in the procedure of the backward problem leads to a more accurate solution in comparison with algorithms that use error minimization. The representation of the backward problem is given in Equation (4).

$$\min_x J(x) \quad s.t. \quad b = Ax \tag{4}$$

where $J(x)$ is the prior knowledge function that is used to promote sparsity. The basis pursuit method uses the l_1-norm to sparsify the problem and transfers it into a linear or quadratic equation (see Equation (5)).

$$\min_x \frac{1}{2}\|b - Ax\|_2^2 + \lambda\|x\|_1 \tag{5}$$

where λ is the scalar regularization parameter, and $\|.\|_1$ and $\|.\|_2$ indicate the l_1-norm and l_2-norm, respectively. The first and second terms of Equation (5) represent the error of estimation and level of sparsity of estimated x, respectively. An effective sparsity method should represent all of the most important features of the PA images. Considering the diagnostically relevant features in a PA image, significant features can be edges, singular points, and homogenous texture.

A standard transform function can be used for signal sparsification. A single transformation, however, can well represent only one of the major features. In order to obtain an image having all these features, we propose combining some of the well-known standard transform functions. Although WT's magnitude will not oscillate around singularities, and it uses continuous transform to characterize the oscillations and discontinuities, the images generated by WT are blurred. DCT helps separate the image into spectral sub-bands of differing importance. In addition, it preserves homogeneous textures better than WT. However, it provides some blocking artifacts in the image. The blocking artifact is a distortion that appears as abnormally large pixel blocks. Therefore, WT and DCT cannot capture two-dimensional singularities, i.e., curves and edges in an image. On the other hand, the TV is an operator that works based on the local variations in a signal. It well preserves the edges in the image. However, the artifacts in the initial image introduced by limited view angles significantly affect the performance of TV [37]. As illustrated, all the three mentioned transformation functions have some benefits and disadvantages. We therefore propose to optimally use the combination of basis functions of WT, DCT, and TV, where WT captures point-like features, DCT captures homogeneous texture components, and TV sharpens the edges and reduces other artifacts without eliminating essential information of the image. In this way, due to the properties of the imaging target, which directly affect the PA reconstructed image generated in the first attempt, the reconstruction procedure is updated, moving toward a high image quality. The procedure of the proposed method can be seen in Figure 1.

The signal (b) can be decomposed to three sub-signals (b_1, b_2, and b_3), each emphasizing a feature in the image, utilizing the morphological component analyses (MCA). Having three sub-signal and three sparse representation methods, MCA assumes that, for each sub-signal, there is a corresponding sparse representation which makes the sub-signal sparser than others. The proposed method is described in Equations (6) and (7), which are the expansion of Equation (5) for the MCA concept.

$$\min_x \frac{1}{2}\|b - (b_1 + b_2 + b_3)\|_2^2 + \alpha\|\psi_w x\|_1 + \beta\|\psi_{DCT} x\|_1 + \gamma TV(x) \tag{6}$$

$$TV(x) = \int |Dx| \tag{7}$$

where x is an initial image, obtained from BP algorithm, and ψ_w and ψ_{DCT} are WT and DCT transform operators, respectively. D is the gradient operator. α, β, and γ are the weight factors for WT, DCT, and TV transform operators, respectively. It should be noted that, in each basis, the emphasis is on one major feature and other features are less signified. These features are preserved as shown in Equation (6).

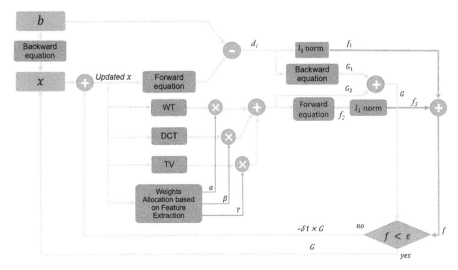

Figure 1. Block diagram of the proposed method. All the italic letters have been described in the text.

Paradigm of the Proposed Method

The proposed reconstruction method, demonstrated in Figure 1, is an iterative algorithm with the raw data acquired by transducer/s as the input and the reconstructed image as the output. The principle of the algorithm is as follows. First, we reconstruct an image (x) using the traditional BP method (considered as the initial image). The image reconstructed by the BP algorithm and the signal used within the BP algorithm (for reconstruction) are considered as the input of the proposed method. Here, b indicates the raw data, detected by the US transducer/s. Using the forward problem, shown in Equation (2), and the initial image (x), we calculate the raw data that could be used to obtain the initial image x. It should be noted that the initial image is obtained by the recorded signals using US transducer/s, but using the forward equation presented in Figure 1, we calculate the signal which could be as the initial signal, ignoring the recorded data. As the next step, we differentiate the recorded signals by the calculated signals. The result of differentiation is called d_1. By applying l_1-norm to d_1, f_1 is generated. The f_1 indicates the power of the error signal. In addition, we calculate an error image, G_1, from the error signal by applying the backward problem illustrated in Equation (3). Simultaneously, the system applies the proposed combinational sparsifying method on x to obtain an image. This image is called G_2, as shown in Figure 1. At this step, the signals that could be used for generation of G_2 is calculated using the forward problem, called f_2. By applying l_1-norm to f_2, f_3 is generated as the power of f_2. Therefore, summation of f_1 and f_3 (i.e., f) is a parameter that shows the difference between the real image and what we have estimated. If f does not meet the desired margin error, x should be updated. The updating process can be done by subtracting a portion of G from x, i.e., δt, iteration step.

Big step accelerates the speed of the program and reduces the execution time, but it may increase the error and lead to an unstable loop. If f meets the desired margin error, the last updated x would be the final image at this first step. The number of the steps is selected based on the trade-off between the execution time and the performance of the final reconstructed image. In our algorithm, we set the number of steps to 50. By allocating appropriate weights to each basis function, coefficients corresponding to dominant features will receive higher gain and those corresponding to less critical features will receive a lower gain.

It should be noted that the weighting factors are determined based on what diagnostically relevant features need to be emphasized in the image, i.e., edges, singular points, and homogeneous textures. A statistical texture analysis method, the Gray-Level Co-Occurrence Matrix (GLCM) technique, has been used to measure the homogeneous texture components [38]. GLCM is used in 0, 45, 90, and 135 degree directions. Singular points are small local areas that the signal values in these areas are changed in two dimensions. The Harris operator, by providing an analytical autocorrelation, considering the local intensity changes in different directions, has been used to detect singular points [39]. We used the canny filter, a first-order image edge detector, to determine the weighting factor TV (which signifies the edges). It should be noted that the main justification for the proposed method is to utilize the advantages of the overlapping methods to improve the image quality in limited view PACT systems. As mentioned, using only one sparsifying function leads to information loss. By combining the three basis, all the information loss in one of them can be mitigated using another one, leading to a higher image quality, compared to using any one of them.

3. Results

To evaluate the performance of the reconstruction algorithms, we made a gel phantom with imaging target inside. The phantom was made by 3% Agar powder in water. Eight straight graphite pencil leads with a 0.5 mm diameter were placed co-centered in the same plane embedded in the gel. The size of the phantom was 30 × 30 mm (Figure 2b). The phantom was imaged using a single-transducer PACT system (Figure 2a). A Quanta-Ray PRO-Series Nd:YAG laser (Spectra-Physics Inc., Santa Clara, CA, USA) pumped a VersaScan V1.7 optical parametric oscillator (OPO) (Spectra-Physics Inc., USA), generating wavelengths in the range of 398–708 nm, with a pulse width of 7 ns and a repetition rate of 30 Hz. In this experiment, the illumination wavelength was 532 nm. A large graded index plastic optical fiber with a diameter of 10 mm and a numerical aperture of 0.55 was used on the top of the ring, 5 cm away from the sample, forming an uniform illumination. A cylindrically focused ultrasound transducer, V326-SU (Olympus Inc., Center Valley, PA, USA) with the element size of 0.375 inch, central frequency of 5 MHz and the focal length of 0.625 inch was positioned on a cylindrical construct. The position of the transducer was in the same plane as the pencil leads were. The diameter of the construct was 75 mm. The cylinder was rotated using a stepper motor (Applied Motion Products Inc., Watsonville, CA, USA) with a driver controlled by LabVIEW. The transducer was smoothly rotated around the phantom with the speed of 0.0125 round/s to collect the PA signal in 360 degrees. The data acquisition is performed using our FPGA based National Instrument (NI) system. Specification of the experiment is presented in Table 1. A photograph of the PAI system is shown in Figure 3. The system was placed on a 12 ft × 4 ft optical table (New Port, Inc., Irvine, CA, USA).

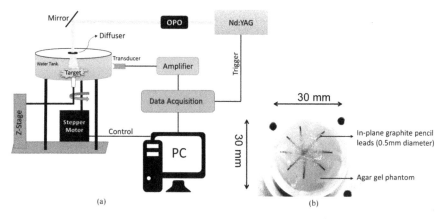

Figure 2. (**a**) Schematic of our PACT system; (**b**) gel phantom with eight 0.5 mm pencil leads.

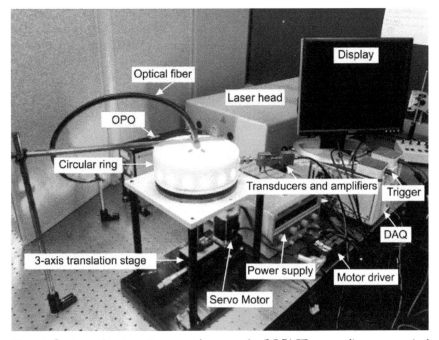

Figure 3. Low-cost photoacoustic computed tomography (LC-PACT) system diagram comprised of an Nd:YAG 30Hz Spectra Physics laser, an optical parametric oscillator (OPO), a circular ring, a DC supply for the motor driver, an NI DAQ, an NI trigger, a servo motor, a motor gear, a three-axis translation stage for phantom, and a transducer-amplifier unit.

Table 1. Specification of the experimental setup presented in Figure 2.

Transducer	5 MHz
Laser energy	20 mJ/cm^2
Laser pulse width	7 ns
Laser rep-rate	30 Hz
Wavelength	532 nm
Amplifier	ZFL500LN
DAQ	National instrument

For image reconstruction, we used a laptop with i7 core, 2.3 GHz CPU, and 8 GB Memory. Figure 4 demonstrates a set of results of different reconstruction algorithms on the phantom data acquired by our PACT system. They are C-scan images with their dimensions annotated in the right image. The reconstructed images with BP and CS with different sparsifying methods for different numbers of views (30, 60, 90, and 120) have been presented. As seen in the first column, BP generates images with a high level of artifacts and imaging noise. Even though increasing the number of angles improves the image quality with lower imaging noise and reconstruction artifacts, the images are still affected. The CS-based WT removes a considerable amount of artifacts presented in the BP image (initial image). However, it blurs the edges of the reconstructed image. Even for a low number of angles, the real structure of the imaging target is compromised. It can be seen that adding the TV sparsifying method to the WT sharpens the edges, especially for 90 and 120 angles. While improvement is obtained, it removes the artifacts in the image along with useful information, lowering the accuracy of the reconstructed image. Finally, the proposed method removes the artifacts effectively and more accurately by sharpening the edges while retaining the significant information in the image. It should be noted that, in all the reconstruction implementations, investigated in this study, the media is assumed to be acoustically homogeneous. It should be noted that some algorithms produce more artifacts and decrease the visibility of the details in reconstructed images. The contrast of the image concerned with the proposed method is not changed, but the amount of artifact is, leading to a better visibility.

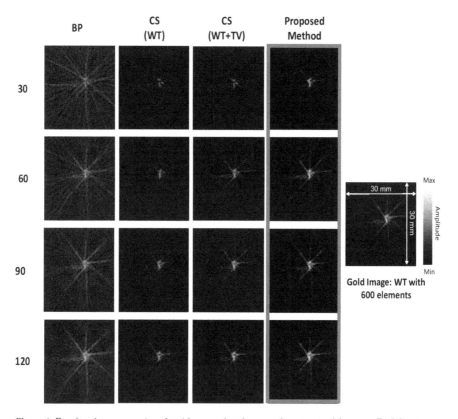

Figure 4. Results of reconstruction algorithms on the phantom data acquired from our PACT system. Different rows show different number of views, 30, 60, 90, and 120, and columns show different reconstruction methods, BP, sparse with basis WT, sparse with basis WT+TV, and sparse with the proposed sparsifying method.

The simulation results performed on Shepp–Logan synthetic phantom with 60 view angles confirm the superiority of the proposed method to other reconstruction algorithms (see Figure 5).

For evaluation of the edge preservation capability of the proposed algorithm, we used the edge preservation index (EPI) [40]. This metric indicates the edge preservation capability in horizontal and perpendicular directions, after applying filters such as the Laplacian operator on the image. The EPI values for different methods/angle views are reported in Figure 6a. Each experiment was repeated 10 times. The values of deviation are shown by error bars in this figure. The value of EPI changes from 0 to 1. A higher value suggests a better ability to preserve the edges. Statistical analysis shows that, for a number of view angles lower than 45, the proposed algorithm outperforms other methods. This indicates that, with lower data acquisition time (a lower number of angles), the proposed method provides a higher image quality, compared to other methods. In addition, Figure 6a shows that the proposed method provides better preserved edges for numbers of angles between 80 and 92. The PSNR is calculated for different methods/angle views, and the results are presented in Figure 6b. PSNR is calculated using the formula presented in [23]. To calculate the minimum square error (MSE), the result of sparse reconstruction with WT basis at the maximum possible number of views, i.e., 600, has been considered as the Gold standard image.

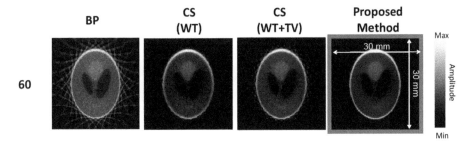

Figure 5. Results of reconstruction algorithms on the Shepp–Logan synthetic phantom for 60 view angles and different reconstruction algorithms.

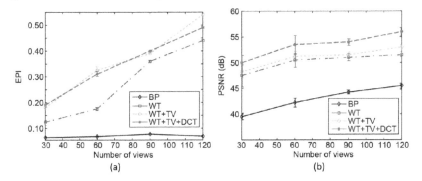

Figure 6. Quantitative evaluation, (a) EPI and (b) PSNR, of the performance of four reconstruction algorithms with different number of view angles.

From the quantitative assessment of the results presented in Figure 6b, it can be seen that, for all numbers of angles, the proposed method results in a higher PSNR. For instance, for 60 angles, BP, WT, WT+TV, and the proposed method lead to PSNR values of about 42.5 dB, 49.5 dB, 50.5 dB, and 54.2 dB, respectively. In other words, PSNR is improved by about 27%, 9%, and 7%, in comparison to BP, WT, and WT+TV, respectively. Considering all the quantitative evaluation, it can be concluded that our proposed algorithm works effectively with a limited number of view angles, outperforming other algorithms. We performed statistical analysis on the PSNR values using SPSS®. At every view angle, the improvement obtained by the proposed method compared to other aforementioned methods was statistically significant (p-value < 0.001).

Table 2 shows the execution time for BP, sparse reconstruction with basis WT, sparse reconstruction with bases WT+TV, and the proposed reconstruction algorithms. The proposed method provides a higher image quality at the expense of a higher processing time, compared to other algorithms. The higher processing time of the proposed algorithm is due to the fact that it should use all the information of the three used sparsifying functions in order to reconstruct a photoacoustic image with a higher image quality. In addition, the feature measurement of the proposed algorithm increases the processing time as well.

Table 2. Execution time for different reconstruction algorithms with 120 view angles.

Algorithm	BP	Sparse (WT)	Sparse (WT+TV)	Proposed
Execution Time (sec)	18.99	387.28	369.79	547.30

4. Conclusions

In this paper, we propose a novel dictionary-based image reconstruction algorithm in which the desired information of the images is shown more accurately. The proposed method uses a combination of WT, DCT, and TV in order to preserve the necessary information within the reconstruction procedure. The algorithm was evaluated experimentally (using a single-element PACT system). EPI and PSNR were used as the quantitative metrics of evaluation. The qualitative and quantitative results presented here show that the proposed algorithm can generate images with specific emphasis on a desired feature, defined by the user. Quantitative analysis of EPI showed that, for a number of angles lower than 45, the proposed algorithm preserves the edges better than other methods. In addition, the calculated PSNR at 60 angles indicated that the proposed method improves PSNR by about 27%, 9%, and 7%, compared with BP, WT, and WT+TV, respectively. At every view angle, the improvement obtained by the proposed method compared to the above-mentioned methods was statistically significant (p-value < 0.001). The proposed technique is particularly useful when a low-cost PACT system (limited number of view angles) with a fine temporal resolution is required. However, all the improvements are obtained at the expense of a higher computational burden. As a feature work, we will evaluate the proposed method in an in vivo study. Accelerating the algorithm as well as reducing its computational complexity is the next major step.

Author Contributions: Conceptualization, P.O. and M.M.; Methodology, P.O., M.M. and A.H.; Software, P.O., A.H., M.Z., M.M. and X.H.; Validation, P.O., M.M., A.H., M.Z. and X.H.; Formal Analysis, P.O., A.H., M.M., X.H.; Investigation, P.O., M.M.; Resources, P.O., M.M., A.H., M.Z. and X.H.; Data Curation, P.O., A.H. M.Z. and X.H.; Writing—Original Draft Preparation, P.O., M.M., A.H., M.Z. and X.H.; Writing—Review & Editing, P.O., M.M., M.N., M.O.; Visualization, M.N. and M.O.; Supervision, M.N. and M.O.; Project Administration, M.N.; Funding Acquisition, M.N.

Funding: This project has been partially supported by a WSU startup fund and the Andersen Institute fund.

Acknowledgments: We thank Jun Xia from Buffalo University for his constructive discussion and help in image reconstruction.

Conflicts of Interest: The authors declare no conflict of interest.

Abbreviations

The following abbreviations are used in this manuscript:

PACT	Photoacoustic Computed Tomography
BP	Back Projection
WT	Wavelet Transform
DCT	Discrete Cosine Transform
TV	Total Variation
EPI	Edge Preservation Index
PSNR	Peak Signal-to-Noise Ratio
FBP	Filtered Back Projection
PA	Photoacoustic
CS	Compressed Sensing
MRI	Magnetic Resonance Imaging
TAI	Theracoustic Imaging
DAQ	Data Acquisition
MCA	Morphological Component Analysis

GLCM   Gray-Level Co-occurrence Matrix
OPO   Optical Parametric Oscillator
NI   National Instrument
MSE   Minimum Square Error

References

1. Xu, M.; Wang, L.V. Photoacoustic imaging in biomedicine. *Rev. Sci. Instrum.* **2006**, *77*, 041101. [CrossRef]
2. Nasiriavanaki, M.; Xia, J.; Wan, H.; Bauer, A.Q.; Culver, J.P.; Wang, L.V. High-resolution photoacoustic tomography of resting-state functional connectivity in the mouse brain. *Proc. Natl. Acad. Sci. USA* **2014**, *111*, 21–26. [CrossRef] [PubMed]
3. Jeon, M.; Song, W.; Huynh, E.; Kim, J.; Kim, J.; Helfield, B.L.; Leung, B.Y.; Geortz, D.E.; Zheng, G.; Oh, J.; et al. Methylene blue microbubbles as a model dual-modality contrast agent for ultrasound and activatable photoacoustic imaging. *J. Biomed. Opt.* **2014**, *19*, 016005. [CrossRef] [PubMed]
4. Mohammadi-Nejad, A.R.; Mahmoudzadeh, M.; Hassanpour, M.S.; Wallois, F.; Muzik, O.; Papadelis, C.; Hansen, A.; Soltanian-Zadeh, H.; Gelovani, J.; Nasiriavanaki, M. Neonatal brain resting-state functional connectivity imaging modalities. *Photoacoustics* **2018**, *10*, 1–19. [CrossRef] [PubMed]
5. Mahmoodkalayeh, S.; Lu, X.; Ansari, M.A.; Li, H.; Nasiriavanaki, M. Optimization of light illumination for photoacoustic computed tomography of human infant brain. In Proceedings of the Photons Plus Ultrasound: Imaging and Sensing 2018, International Society for Optics and Photonics, San Francisco, CA, USA, 27 January–1 February 2018; Volume 10494, p. 104946U.
6. Mozaffarzadeh, M.; Mahloojifar, A.; Orooji, M.; Kratkiewicz, K.; Adabi, S.; Nasiriavanaki, M. Linear-array photoacoustic imaging using minimum variance-based delay multiply and sum adaptive beamforming algorithm. *J. Biomed. Opt.* **2018**, *23*, 026002. [CrossRef] [PubMed]
7. Hariri, A.; Fatima, A.; Mohammadian, N.; Mahmoodkalayeh, S.; Ansari, M.A.; Bely, N.; Avanaki, M.R. Development of low-cost photoacoustic imaging systems using very low-energy pulsed laser diodes. *J. Biomed. Opt.* **2017**, *22*, 075001. [CrossRef] [PubMed]
8. Mozaffarzadeh, M.; Mahloojifar, A.; Orooji, M.; Adabi, S.; Nasiriavanaki, M. Double-Stage Delay Multiply and Sum Beamforming Algorithm: Application to Linear-Array Photoacoustic Imaging. *IEEE Trans. Biomed. Eng.* **2018**, *65*, 31–42. [CrossRef] [PubMed]
9. Mozaffarzadeh, M.; Mahloojifar, A.; Periyasamy, V.; Pramanik, M.; Orooji, M. Eigenspace-Based Minimum Variance Combined with Delay Multiply and Sum Beamformer: Application to Linear-Array Photoacoustic Imaging. *IEEE J. Sel. Top. Quantum Electron.* **2019**, *25*, 1–8. [CrossRef]
10. Yeh, C.; Li, L.; Zhu, L.; Xia, J.; Li, C.; Chen, W.; Garcia-Uribe, A.; Maslov, K.I.; Wang, L.V. Dry coupling for whole-body small-animal photoacoustic computed tomography. *J. Biomed. Opt.* **2017**, *22*, 041017. [CrossRef] [PubMed]
11. Cho, Y.; Chang, C.C.; Yu, J.; Jeon, M.; Kim, C.; Wang, L.V.; Zou, J. Handheld photoacoustic tomography probe built using optical-fiber parallel acoustic delay lines. *J. Biomed. Opt.* **2014**, *19*, 086007. [CrossRef] [PubMed]
12. Li, G.; Li, L.; Zhu, L.; Xia, J.; Wang, L.V. Multiview Hilbert transformation for full-view photoacoustic computed tomography using a linear array. *J. Biomed. Opt.* **2015**, *20*, 066010. [CrossRef] [PubMed]
13. Zhang, P.; Li, L.; Lin, L.; Hu, P.; Shi, J.; He, Y.; Zhu, L.; Zhou, Y.; Wang, L.V. High-resolution deep functional imaging of the whole mouse brain by photoacoustic computed tomography in vivo. *J. Biophotonics* **2018**, *11*, e201700024. [CrossRef] [PubMed]
14. Jo, J.; Tian, C.; Xu, G.; Sarazin, J.; Schiopu, E.; Gandikota, G.; Wang, X. Photoacoustic tomography for human musculoskeletal imaging and inflammatory arthritis detection. *Photoacoustics* **2018**, in press. [CrossRef]
15. Lin, L.; Xia, J.; Wong, T.T.; Li, L.; Wang, L.V. In vivo deep brain imaging of rats using oral-cavity illuminated photoacoustic computed tomography. *J. Biomed. Opt.* **2015**, *20*, 016019. [CrossRef] [PubMed]
16. Wang, J.; Wang, Y. An Efficient Compensation Method for Limited-View Photoacoustic Imaging Reconstruction Based on Gerchberg–Papoulis Extrapolation. *Appl. Sci.* **2017**, *7*, 505. [CrossRef]

17.   Li, L.; Zhu, L.; Ma, C.; Lin, L.; Yao, J.; Wang, L.; Maslov, K.; Zhang, R.; Chen, W.; Shi, J.; et al. Single-impulse panoramic photoacoustic computed tomography of small-animal whole-body dynamics at high spatiotemporal resolution. *Nat. Biomed. Eng.* **2017**, *1*, 0071. [CrossRef] [PubMed]

18.   Matthews, T.P.; Anastasio, M.A. Joint reconstruction of the initial pressure and speed of sound distributions from combined photoacoustic and ultrasound tomography measurements. *Inverse Probl.* **2017**, *33*, 124002. [CrossRef] [PubMed]

19.   Matthews, T.P.; Wang, K.; Li, C.; Duric, N.; Anastasio, M.A. Regularized dual averaging image reconstruction for full-wave ultrasound computed tomography. *IEEE Trans. Ultrason. Ferroelectr. Freq. Control* **2017**, *64*, 811–825. [CrossRef] [PubMed]

20.   Xu, M.; Wang, L.V. Universal back-projection algorithm for photoacoustic computed tomography. *Phys. Rev. E* **2005**, *71*, 016706. [CrossRef] [PubMed]

21.   Egolf, D.M.; Chee, R.K.; Zemp, R.J. Sparsity-based reconstruction for super-resolved limited-view photoacoustic computed tomography deep in a scattering medium. *Opt. Lett.* **2018**, *43*, 2221–2224. [CrossRef] [PubMed]

22.   Dean-Ben, X.L.; Buehler, A.; Ntziachristos, V.; Razansky, D. Accurate model-based reconstruction algorithm for three-dimensional optoacoustic tomography. *IEEE Trans. Med. Imaging* **2012**, *31*, 1922–1928. [CrossRef] [PubMed]

23.   Zhang, Y.; Wang, Y.; Zhang, C. Total variation based gradient descent algorithm for sparse-view photoacoustic image reconstruction. *Ultrasonics* **2012**, *52*, 1046–1055. [CrossRef] [PubMed]

24.   Zhang, Y.; Wang, Y.; Zhang, C. Efficient discrete cosine transform model–based algorithm for photoacoustic image reconstruction. *J. Biomed. Opt.* **2013**, *18*, 066008. [CrossRef] [PubMed]

25.   Zhang, C.; Zhang, Y.; Wang, Y. A photoacoustic image reconstruction method using total variation and nonconvex optimization. *Biomed. Eng. Online* **2014**, *13*, 117. [CrossRef] [PubMed]

26.   Mozaffarzadeh, M.; Mahloojifar, A.; Nasiriavanaki, M.; Orooji, M. Model-based photoacoustic image reconstruction using compressed sensing and smoothed L0 norm. In Proceedings of the Photons Plus Ultrasound: Imaging and Sensing 2018, International Society for Optics and Photonics, San Francisco, CA, USA, 27 January–1 February 2018; Volume 10494, p. 104943Z.

27.   Guo, Z.; Li, C.; Song, L.; Wang, L.V. Compressed sensing in photoacoustic tomography in vivo. *J. Biomed. Opt.* **2010**, *15*, 021311. [CrossRef] [PubMed]

28.   Provost, J.; Lesage, F. The application of compressed sensing for photo-acoustic tomography. *IEEE Trans. Med. Imaging* **2009**, *28*, 585–594. [CrossRef] [PubMed]

29.   Rosenthal, A.; Razansky, D.; Ntziachristos, V. Quantitative optoacoustic signal extraction using sparse signal representation. *IEEE Trans. Med. Imaging* **2009**, *28*, 1997–2006. [CrossRef] [PubMed]

30.   Donoho, D.L. Compressed sensing. *IEEE Trans. Inf. Theory* **2006**, *52*, 1289–1306. [CrossRef]

31.   Ramirez-Giraldo, J.; Trzasko, J.; Leng, S.; Yu, L.; Manduca, A.; McCollough, C.H. Nonconvex prior image constrained compressed sensing (NCPICCS): Theory and simulations on perfusion CT. *Med. Phys.* **2011**, *38*, 2157–2167. [CrossRef] [PubMed]

32.   Xi, Y.; Zhao, J.; Bennett, J.R.; Stacy, M.R.; Sinusas, A.J.; Wang, G. Simultaneous CT-MRI reconstruction for constrained imaging geometries using structural coupling and compressive sensing. *IEEE Trans. Biomed. Eng.* **2016**, *63*, 1301–1309. [CrossRef] [PubMed]

33.   Qin, T.; Wang, X.; Qin, Y.; Wan, G.; Witte, R.S.; Xin, H. Quality improvement of thermoacoustic imaging based on compressive sensing. *IEEE Antennas Wirel. Propag. Lett.* **2015**, *14*, 1200–1203. [CrossRef]

34.   Meng, J.; Wang, L.V.; Liang, D.; Song, L. In vivo optical-resolution photoacoustic computed tomography with compressed sensing. *Opt. Lett.* **2012**, *37*, 4573–4575. [CrossRef] [PubMed]

35.   Liu, Q.; Wang, S.; Ying, L.; Peng, X.; Zhu, Y.; Liang, D. Adaptive dictionary learning in sparse gradient domain for image recovery. *IEEE Trans. Image Process.* **2013**, *22*, 4652–4663. [CrossRef] [PubMed]

36.   Zhou, W.; Cai, J.F.; Gao, H. Adaptive tight frame based medical image reconstruction: a proof-of-concept study for computed tomography. *Inverse Probl.* **2013**, *29*, 125006. [CrossRef]

37.   Duarte-Carvajalino, J.M.; Sapiro, G. Learning to sense sparse signals: Simultaneous sensing matrix and sparsifying dictionary optimization. *IEEE Trans. Image Process.* **2009**, *18*, 1395–1408. [CrossRef] [PubMed]

38.   Haralick, R.M. Statistical and structural approaches to texture. *Proc. IEEE* **1979**, *67*, 786–804. [CrossRef]
39.   Harris, C.; Stephens, M. A combined corner and edge detector. In Proceedings of the Alvey Vision Conference, Manchester, UK, 31 August–2 September 1988; Volume 15, pp. 10–5244.
40.   Adabi, S.; Mohebbikarkhoran, H.; Mehregan, D.; Conforto, S.; Nasiriavanaki, M. An intelligent despeckling method for swept source optical coherence tomography images of skin. In Proceedings of the Medical Imaging 2017: Biomedical Applications in Molecular, Structural, and Functional Imaging, International Society for Optics and Photonics, Orlando, FL, USA, 13 March 2017; Volume 10137, p. 101372B.

Article

A Single Simulation Platform for Hybrid Photoacoustic and RF-Acoustic Computed Tomography

Christopher Fadden [1] and Sri-Rajasekhar Kothapalli [2,3,*]

[1] Department of Electrical Engineering, The Pennsylvania State University, University Park, PA 16802, USA; czf41@psu.edu

[2] Department of Biomedical Engineering, The Pennsylvania State University, University Park, PA 16802, USA

[3] Penn State Hershey Cancer Institute, The Pennsylvania State University, Hershey, PA 17033, USA

* Correspondence: szk416@engr.psu.edu

Received: 3 July 2018; Accepted: 29 August 2018; Published: 6 September 2018

Abstract: In recent years, multimodal thermoacoustic imaging has demonstrated superior imaging quality compared to other emerging modalities. It provides functional and molecular information, arising due to electromagnetic absorption contrast, at ultrasonic resolution using inexpensive and non-ionizing imaging methods. The development of optical- as well as radio frequency (RF)-induced thermoacoustic imaging systems would benefit from reliable numerical simulations. To date, most numerical models use a combination of different software in order to model the hybrid thermoacoustic phenomenon. Here, we demonstrate the use of a single open source finite element software platform (ONELAB) for photo- and RF-acoustic computed tomography. The solutions of the optical diffusion equation, frequency domain Maxwell's equations, and time-domain wave equation are used to solve the optical, electromagnetic, and acoustic propagation problems, respectively, in ONELAB. The results on a test homogeneous phantom and an approximate breast phantom confirm that ONELAB is a very effective software for both photo- and RF-acoustic simulations, and invaluable for developing new reconstruction algorithms and hardware systems.

Keywords: photoacoustic imaging; tomography; thermoacoustic; radio frequency

1. Introduction

One of the goals of modern medical imaging is to simultaneously provide molecular, functional, and structural/anatomical information corresponding to various tissues. To achieve such comprehensive information, a combination of conventional imaging technologies such as X-ray computed tomography (CT) [1], positron emission tomography (PET) [2] and magnetic resonance imaging (MRI) [3] are used. CT provides anatomical contrast, PET provides molecular and metabolic information, and MRI provides both functional and anatomical contrasts. A PET–CT or PET–MRI combination is therefore widely used for simultaneously mapping molecular and anatomical contrasts [4–6].

All of these methods have significant downsides: one of which is the financial burden, which means that these methods are not ideal for routine imaging, and a second is the ionizing radiation used in CT and PET. As an alternative, ultrasound imaging that uses non-ionizing radiation is routinely used in several clinical applications for anatomical imaging [7]. However, it lacks the molecular or functional information necessary for detecting the early symptoms of disease.

Alternative medical imaging modalities that provide anatomical, functional, and molecular information about the tissue, while being lower cost and not being as restrictive to patient movement, are needed. Photoacoustic computed tomography (PACT) and radio frequency (RF)-induced acoustic computed tomography (RACT), together known under the general term thermoacoustic computed tomography (TACT), match these objectives of lower cost functional/molecular imaging, using non-ionizing electromagnetic radiation [8–16]. While PACT maps optical absorption contrast using optical radiation induced acoustic wave detection, RACT maps tissue conductivity using RF induced acoustic wave detection. More importantly, since these hybrid (combining electromagnetic radiation and acoustic detection) imaging modalities share the same ultrasound detection platform, combined trimodality PACT–RACT–UCT (ultrasound computed tomography) systems have been realized for mapping functional, molecular, and anatomical contrasts [17]. The molecular absorption of electromagnetic energy causes thermal expansion in the tissue, which then leads to generation of acoustic waves. The acoustic waves propagate out of the tissue and are received by ultrasound transducers located at the boundary of the body. This data is then used to reconstruct thermoacoustic images displaying electromagnetic absorption contrast at ultrasonic spatial resolution. The imaging depth and spatial resolution in TACT is scalable with the frequency of excitation radiation and ultrasound transducer. The fact that the detected acoustic signal arises directly from specific molecules inside the tissue makes TACT a molecular/functional imaging technology. In PACT, the tissue chromophores—such as oxy-hemoglobin, deoxy-hemoglobin, melanin, and lipids—absorb light photons in the wavelength range from 400 nm to 1200 nm. By using different wavelengths, which takes advantage of the resonance peaks in the absorption spectrum of the imaged molecules, the distribution of different molecules inside the tissue can be mapped. In RACT, a radio frequency source in the frequency range from 434 MHz to 9 GHz is used for mapping the water distribution. The thermoacoustic effect in this frequency range is dependent on the conductivity distribution of the medium. The conductivity difference between water, tissue, and tumors can then give useful functional images.

Accurate numerical modeling of TACT is paramount for the development of robust reconstruction algorithms to quantify the electromagnetic absorption properties of the tissue. In PACT, the forward optical simulation of total light fluence (Φ), calculated inside the tissue medium, is usually achieved using either Monte Carlo simulations or the software package NIRFast, which solves the light diffusion equation. The fluence distribution is then converted to an initial pressure rise, which is further propagated through the tissue medium and detected by the ultrasound transducers located on the boundary using acoustic simulation tools such as the K-wave toolbox. NIRFast [18] uses the finite element method (FEM), while K-wave uses a spectral-based finite difference method (FDM) to model the propagation of acoustic waves [19]. The goal of the reconstruction problem is then to recover the tissue properties (optical absorption in PACT and conductivity in RACT) given only the sensor data.

Groups that use a single simulation platform for both the optical and acoustic propagation needed for the hybrid PACT/RACT technique are less common. Recently, there are studies that use the commercial software COMSOL to solve both propagation problems with a single software package [20,21]. As an alternative to the commercial software, the simulation system described in this paper uses the open source softwares Gmsh [22] and GetDP [23], often combined under the name ONELAB [24]. ONELAB is an FEM solver, which uses Gmsh for creating the FEM mesh, and GetDP for solving generic partial differential equations (PDEs) with the FEM method. Advantages of using Gmsh include its ability to create user defined meshes, but also having standard interfaces with other commonly used mesh and computer-aided design (CAD) software such as STEP, IGES, and STL. Segmented DICOM images commonly used in MRI can then be converted to a mesh that Gmsh understands, creating realistic phantoms on which to test algorithms. The ability of GetDP to solve generic PDEs allows the user to implement algorithms to solve for optical and acoustic propagation, similar to the combination of NIRFast and K-wave. Both propagation methods, as well as any reconstruction methods, are implemented in the same mesh, with no loss of precision. Since there is less chance of numerical error, a single software platform represents a more accurate approximation of a real-world scenario. Besides PACT with optical sources, we demonstrate that the generality of GetDP allows for sources in the RF regime of the electromagnetic spectrum to also be simulated for RACT. The use of the ONELAB software package allows accurate TACT modeling, in order to develop algorithms for functional imaging of the human body. Although there are experimental studies combining PACT and RACT techniques into one setup, perhaps our study is the first to report a single TACT simulation platform for simulating both PACT and RACT.

The rest of the paper is organized as follows: In Section 2, we describe methods and materials. Sections 3.1 and 3.2 show PACT and RACT results on a homogeneous phantom. Sections 3.3 and 3.4 show PACT and RACT results on an approximate breast phantom. Section 4 is a discussion of the results in Section 3, comparing errors in reconstruction between PACT and RACT, and between the two phantoms. Section 5 concludes the paper.

2. Materials and Methods

There are three main phenomena that need to be modeled in thermoacoustics. The first is the propagation of the initial energy source, in this study near infrared optical or radio frequency electromagnetic waves, and calculation of the total fluence/intensity distribution inside the tissue medium. The second phenomenon is the acoustic propagation, after calculating an initial pressure from the intensity distribution. The third phenomenon is the reconstruction of the tissue parameters, which depend on the type of energy source used for tissue excitation. A flowchart showing the similar steps between RACT and PACT is shown in Figure 1. This study is focused on using a time reversal [25,26] reconstruction method to recover the initial pressure, followed by a simple division to find the tissue parameters. There are several alternative approaches for reconstructing the tissue parameters that can be integrated with our study in the future. Back-projection methods may be used when using specific geometries [27] and parametrix methods for more general geometries [28]. The back-projection methods are analytically exact, but only for geometries with a special symmetry, and with a constant speed of sound. Parametrix methods do not provide analytically exact methods, but instead give approximations with error bounds for general geometries, and can even be developed for regions with varying speeds of sound. A survey of several reconstruction methods is given in Reference [29].

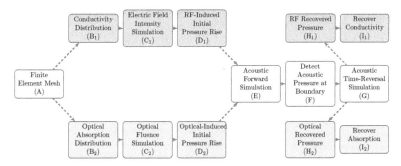

Figure 1. The similarities between key steps of the photoacoustic computed tomography (PACT, blue) and radio frequency (RF)-induced acoustic computed tomography (RACT, red) workflows, with methods that are shared represented in yellow. Step A represents the finite element mesh in Section 2.5, used for both types of simulation. Step B_1 is represented in Figure 9a, while step B_2 is represented in Figure 7a. Step C_2 uses the optical diffusion Equation (1). Step C_1 uses the electromagnetic wave Equation (5). The resulting pressure rise for the optical (D_2) and RF (D_1) radiations are represented by Figures 7b and 9b, respectively. The initial pressure propagates out to the boundary (E) via the scalar wave Equation (7). The data received by acoustic sensors at the boundary (F) can then be used in reconstruction algorithms (G) to recover the initial pressure. In this paper, a time-reversal algorithm is applied for reconstruction (Section 2.4). The reconstructed pressure is then divided by the intensity (I_2, I_1) to recover the recover the absorption or conductivity for the optical and RF cases, shown in Figures 7c and 9c, respectively.

2.1. Optical Propagation for Photoacoustic Computed Tomography (PACT)

A full description of optical propagation would use the radiative transport equation in its full generality. This equation can be modeled by the use of Monte Carlo methods, with significant computational cost [30,31]. Since the the photoacoustic effect depends on the optical fluence rate, and the wavelengths used correspond to the near infrared (NIR) regime, a diffusion approximation to the radiative transport equation is often used [32]. The diffusion equation for the fluence rate is:

$$- \nabla \cdot \kappa(x) \nabla \Phi(x, \lambda) + \mu_a(x) \Phi(x) = q(x) \ x \in \Omega \tag{1}$$

$$\Phi(x) = 0 \ x \in \partial \Omega. \tag{2}$$

In this equation, Φ is the optical fluence rate, or intensity of the light. μ_a is the absorption coefficient at each point in space, while $\kappa = \frac{1}{3(\mu_a + \mu_s')}$ is the diffusion coefficient, calculated using the absorption coefficient and the reduced scattering coefficient μ_s'. q is the source function, the initial laser pulse that irradiates the tissue medium. Ω is the tissue domain of interest, usually a subset of $\mathbb{R}^n$, with n corresponding to the dimension of the problem (2 or 3). Equation (2) corresponds to the Dirichlet boundary condition used in this study. The amount of pressure generated by the optical energy is given by a constant of proportionality, the Gruneisen parameter $\Gamma = \frac{\alpha v_s^2}{C_p}$, where α is the volume thermal expansion coefficient, v_s is the speed of sound, and C_p is the heat capacity at constant pressure [33]. Therefore, for a given fluence rate Φ, the initial pressure is given by:

$$p_0(x) = \Gamma(x) \Phi(x) \mu_a(x). \tag{3}$$

While Γ will physically vary over space, this small deviation is ignored in this study, and the parameter is assumed constant, with a value of 0.1, which approximates standard tissue.

2.2. Radio Frequency Propagation for RF Acoustic Computed Tomography (RACT)

Radio frequency radiation is modeled by Maxwell's equations, which when modeled at a single frequency can be reduced to the following wave equation:

$$\nabla \times \nabla \times E(x) - \omega^2 \mu(x) \varepsilon(x) E(x) = q(x) \ x \in \Omega. \tag{4}$$

In this equation, E is the complex electric field, $\omega = 2\pi f$ is the frequency, and μ and ε are the permeability and permittivity of the medium, respectively. In a material with electric loss, the permittivity is complex valued, with $\varepsilon = \varepsilon_{real} - j\frac{\sigma}{\omega}$, and $\sigma = 0$ for a medium with no loss. When the simulation is restricted to two dimensions (assuming a TM_z polarization), Equation (4) can be reduced to a scalar Helmholtz equation:

$$\nabla^2 E_3(x) + k(x)^2 E_3(x) = q(x) \ x \in \Omega, \tag{5}$$

where $k = \frac{\omega}{c}$ is the wavenumber, and can be complex in a tissue with energy loss. For boundary conditions, a Dirichlet boundary condition similar to Equation (2) is implemented. However, this would not prevent the electromagnetic waves from reflecting off the boundaries of the domain, causing numerical artifacts. Therefore, while there is a Dirichlet boundary condition, there is also a non-physical space occupied by a perfectly matched layer (PML) to absorb the outgoing waves and prevent reflection [34]. Once the electric field is calculated, the initial pressure is given by three contributions [35]. The first contribution is due to the conductivity of the medium, $p_{cond} = \int_V \frac{\sigma}{2}|E|^2 dV$. The second due to the permittivity $\int_V \frac{\varepsilon_0 \varepsilon_r}{2}|E|^2 dV$, and the third the permeability $\int_V \frac{\mu_0 \mu_r}{2}|H|^2 dV$. In practice, for the tissue media of interest in thermoacoustic tomography, the contribution to the pressure due to the conductivity dominates over the permittivity and permeability, and so these terms are generally ignored. Therefore, similar to the optical absorption in Equation (3), the initial pressure due to the electric field is given as:

$$p_0(x) = \Gamma(x)\frac{\sigma(x)}{2}|E(x)|^2. \tag{6}$$

2.3. Thermoacoustic Equation

Once an initial pressure is found, via either optical or radio frequency electromagnetic radiation, the pressure must then propagate to ultrasound transducers located outside the tissue boundary. Extremely short (nanosecond) pulses of energy are assumed to be irradiating the medium. Therefore the initial pressures are assumed to be delta functions, and only present for the initial conditions when formulating the equations. The pressure wave in photoacoustics is usually modelled using a standard scalar wave equation, with initial conditions:

$$\frac{\partial^2}{\partial t^2}p(x,t) = c^2(x)\nabla^2 p(x,t) \quad x \in \Omega$$
$$p(x,0) = f(x) \tag{7}$$
$$\frac{\partial}{\partial t}p(x,0) = 0,$$

where, $f(x)$ corresponds to the initial pressure found from the initial pulsed excitation, and $p(x,t)$ is the pressure wave that propagates to the transducers. Similar to the electromagnetic wave equation, the pressure propagation also requires a perfectly matched layer (PML) to prevent reflections from the boundary of the numerical domain [36,37]. Therefore, the pressure is received at transducers along a curve (surface in 3D) γ, which is not the boundary of the domain $\partial\Omega$.

2.4. Time Reversal Reconstruction Algorithm

The pressure received at the transducers can be represented as $g(y,t)$, with $y \in \gamma$. The goal of the reconstruction problem is to reconstruct the tissue parameters, μ_a in the optical excitation or σ in the RF excitation case, given $g(y,t)$. One way of recovering the parameters is to try and find the initial pressure $f(x)$ from the pressure measurements $g(y,t)$, and then divide the pressure by the assumed known energy distribution to recover the parameters. One of the more common methods to recover the initial pressure is using the time-reversal technique. For the time-reversal to rigorously reconstruct $f(x)$, requires Huygen's principle to be valid. Unfortunately, Huygen's principle does not hold when the speed of sound is not constant, and more problematically, it does not hold in two dimensions [38]. Therefore, in most domains of interest the time-reversal method recovers an approximation of the initial pressure $f(x)$.

As the name suggests, time reversal entails taking the received pressure $g(y,t)$, and using it as a source on γ, with time moving in reverse. If the forward simulation was run until a stoppage time T, then when simulated using the time reversal method $p(x,T) \approx f(x)$. There are several other methods that can be used to recover $f(x)$ besides time-reversal, but there is no universally accepted reconstruction algorithm that works in all cases of interest. Most of these methods require pressure information at the sensors as functions of time. For this reason, the computationally expensive time-domain wave equation was used for solving the pressure wave equation instead of a Helmholtz equation similar to Equation (5). Since the initial pressure acts as a delta function, the received signal is inherently broadband. A time domain simulation is more efficient for measuring broadband response, and provides a better representation of what would be measured during a physical experiment.

2.5. Phantom Geometry

Two phantoms are used to demonstrate the effectiveness of the single simulation tool used in this study. The first homogeneous phantom consists of a circular region of interest with radius 20 mm. Within this region, two objects are placed, a circle of radius 1 mm placed approximately 4 mm to the left of center, and an ellipse with major axis 2 mm and minor axis 1 mm placed 4 mm to the right of center. The second phantom consists of an approximation to a human breast. The breast region of interest is represented by a circle of radius 50 mm. Glandular tissue rendered as a circle of radius 10 mm surrounds an elliptical tumor with major axis 7 mm and minor axis 3.5 mm. The tumor is located 32.5 mm deep from the right side of the phantom. The optical sources and acoustic detectors are placed around the region of interest in a continuous fashion.

For creating the finite element meshes, a characteristic length of 2 mm was used for efficiency reasons, though near the absorbing objects, the size of the elements smoothly decreased in order to properly model the objects with sufficient resolution. For the homogeneous phantom mesh, the total number of nodes in the mesh was 3635 with 7268 elements. The mesh for the breast phantom had 12,166 nodes and 24,330 elements. For the time domain simulation of the acoustic propagation, a time step of 30 nanoseconds was used in a Newmark numerical integration method. The homogeneous phantom ran for 700 time steps, while the breast phantom ran for 1700 time steps. A constant speed of sound of 1.5 mm/µs was used for both the forward and time-reversed simulations.

The time estimates for the breast phantom are: 0.6 s for the optical simulation, and 1.1 s for the RF simulation. The acoustic simulation took 18.1 s in the optical source case, and 17.3 s in the RF source case. The reconstruction using time reversal took 385.8 s and 408.0 s for the optical and RF source cases, respectively. The homogeneous phantom timings are: 0.16 s and 0.315 s for the optical and RF simulations; 2.1 s and 2.0 s for the forward acoustic simulation; and the time reversal took 38.4 s and 39.4 s for the optical and RF cases, respectively. The timings were done on a Dell Precision 5820 desktop PC, on a single thread. ONELAB has the functionality to run on multiple threads, as well as a graphical processing unit (GPU), but these options were not used in measuring these time estimates.

2.6. Phantom Parameters

The absorption coefficient μ_a of the homogeneous phantom for the background was set to 0.001 mm^{-1}. The circle and ellipse, which represent absorbers, have an absorption coefficient of 0.425 mm^{-1}; the average absorption coefficient of blood at 800 nm. As is common in human tissue, the background is assumed to have a higher reduced scattering coefficient μ_s' than the absorbers. The reduced scattering coefficient for the background is set to 1 mm^{-1}, while the absorbers μ_s' is set to zero for this phantom. In terms of electrical properties, the human body in general does not have a significant magnetic response at radio frequencies, and so the relative permeability μ_r is set to 1 for all objects. A background relative permittivity of $\varepsilon_r = 5$ is similar to human tissue at 434 MHz. Unlike in the PACT case, objects such as tumors have a higher scattering coefficient as well as absorption in the radio frequency regime. Therefore, the relative permittivity of the circle and ellipse were set to $\varepsilon_r = 25$. The conductivity, which governs the amount of absorption, was set to 0.1 S/m for the background, approximating general tissue. The circle and ellipse are given values similar to that of a tumor, 10 S/m [39].

General breast tissue has an absorption coefficient $\mu_a = 0.005$ mm^{-1}, with a reduced scattering coefficient of 1.52 mm^{-1} when an 800 nm source is used. The optical properties of the glandular tissue and tumor depend on the assumed material composition of the tissues. The assumed amount of hemoglobin and percentage of water can have a dramatic effect on the properties at any given wavelength. In this study, the same composition as Reference [18] is used for the glandular and tumor tissue. The spectral characteristics of hemoglobin, deoxyhemoglobin, and water at 800 nm are taken from References [40,41]. The specific optical properties used for this breast phantom are given in Table 1.

Table 1. Optical properties (μ_a the absorption coefficient, and μ_s' the reduced scattering coefficient) of different breast tissue at 800 nm.

Tissue	μ_a (mm^{-1})	μ_s' (mm^{-1})
Background	0.0005	1.5742
Glandular	0.0059	1.12
Tumor	0.0021	0.625

Human tissue is dispersive at radio frequencies, and so the electrical properties vary over frequency, though not as dramatically as the optical parameters. The properties for the generic and glandular tissue at 434 MHz were taken from the ITIS database provided by ETH-Zurich [42], which references a technical report compiled by the United States Air Force [43,44]. The tumor properties were extrapolated from data provided at 100 MHz in Reference [39] using a Cole–Cole dispersion model. The specific electromagnetic parameters are given in Table 2. The relative magnetic permeability is again set to unity, since the body does not exhibit strong magnetic response at 434 MHz.

Table 2. Electrical properties (ε_r the relative permittivity, and σ the conductivity) of different breast tissue at 434 MHz.

Tissue	ε_r	σ (S/m)
Background	5.51	0.0353
Glandular	61.3	8.86
Tumor	25.25	13.03

3. Results

The ability of the ONELAB software platform to simulate both PACT (optical source) and RACT (radio frequency source) with the same tool and on the same mesh, is demonstrated using the phantoms described in the methods section. Below we first present PACT and RACT results for the homogeneous phantom embedded with two absorbers and then the results for the breast phantom with tumor.

3.1. Photoacoustic Computed Tomography (PACT) of the Homogeneous Phantom

The optical excitation for the first phantom uses a wavelength of 800 nm. This specific wavelength was chosen as it is often used in photoacoustics, due to the absorption spectrum of the absorbers of interest, such as hemoglobin. With the homogeneous phantom parameters, the simulated optical fluence is given in Figure 2.

The position of the absorbers is easily approximated by the nulls in the fluence distribution. The approximate shapes can be identified, which can be used for more stable reconstruction of the material parameters. The 8 mm spacing is large enough to identify two distinct absorbers, with the ultimate resolution governed by the reconstruction of the fluence that would be done in practice. The initial pressure induced by this fluence distribution, as well as the original and reconstructed absorption coefficient images are provided in Figure 3.

The fluence is not constant across the domain, and so the two absorbers induce slightly different initial pressures. The larger object is reconstructed with less error, since it is approximately constant over a larger area. Non-idealities in the reconstructed background pressure are suppressed by using the fluence, and so the reconstructed objects are clearly separated from the background.

3.2. RF-Acoustic Computed Tomography (RACT) of the Homogeneous Phantom

The RACT simulation used the same finite element mesh that was used for the photoacoustic simulation, utilizing a typically used radio frequency source of 434 MHz. For the parameters given in the methods section, the electric field magnitude is provided in Figure 4.

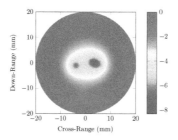

Figure 2. Total optical fluence distribution inside the homogeneous phantom with two absorbers using an 800 nm source. The circular and elliptical absorbers, simulating hemoglobin with absorption coefficient $\mu_a = 0.425$ mm^{-1} at 800 nm, are located 4 mm to the left and right of the center. The background has a $\mu_a = 0.001$ mm^{-1}, and a reduced scattering coefficient $\mu'_s = 1$ mm^{-1}. $\mu'_s = 0$ for the absorbers.

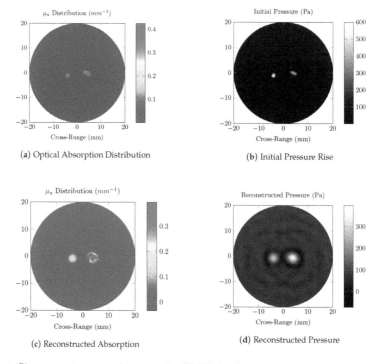

(**a**) Optical Absorption Distribution

(**b**) Initial Pressure Rise

(**c**) Reconstructed Absorption

(**d**) Reconstructed Pressure

Figure 3. Photoacoustic computed tomography (PACT) simulations using an 800 nm source on the homogeneous phantom with two light absorbing inclusions; (**a**) the true optical absorption distribution; (**b**) the initial pressure rise induced by the photoacoustic effect; (**d**) the reconstructed pressure, calculated using the time-reversal algorithm; and (**c**) the reconstructed absorption, obtained by dividing the reconstructed pressure with the fluence.

Unlike the PACT case, the electric field alone does not provide any information on the location of the absorbers. Since the wavelength is much larger than the simulation domain, and there is small difference in conductivity between the absorbers and background, the electric field intensity is approximately constant. The results of the RF acoustic simulation of this homogeneous phantom are shown in Figure 5.

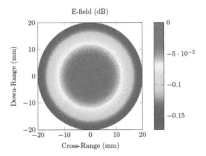

Figure 4. Total electric field intensity distribution inside the homogeneous phantom containing two absorbers, using a 434 MHz radio frequency (RF) source. The circular and elliptical absorbers, with conductivity $\sigma = 10\,\text{S/m}$, are located 4 mm to the left and right of the center, and have a dielectric constant $\varepsilon_r = 25$. The background has $\sigma = 0.1\,\text{S/m}$, and $\varepsilon_r = 5$.

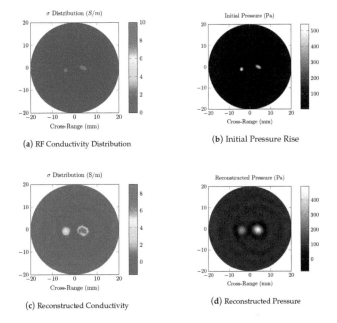

(a) RF Conductivity Distribution

(b) Initial Pressure Rise

(c) Reconstructed Conductivity

(d) Reconstructed Pressure

Figure 5. Radio frequency (RF)-induced acoustic computed tomography (RACT) simulations using a 434 MHz RF source on the homogeneous phantom with two absorbers. (**a**) The true RF conductivity distribution; (**b**) the initial pressure rise induced by the thermoacoustic effect; (**d**) the reconstructed pressure, calculated using the time-reversal algorithm; and (**c**) the reconstructed conductivity, obtained by dividing the reconstructed pressure by the electric field intensity.

3.3. Photoacoustic Computed Tomography (PACT) of the Breast Phantom

Figure 6 shows the fluence distribution generated by ONELAB for the breast phantom described in the methods section. A rough estimate of the tumor location can be predicted from the fluence map, but no shape information can be obtained from the fluence distribution alone. The photoacoustic simulation is able to identify the shape of the tumor, as well as the difference between cancerous and glandular tissue. Results in Figure 7 show the initial pressure rise and the resulting reconstructed absorption distribution.

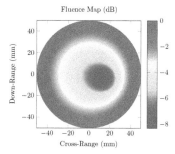

Figure 6. Total optical fluence distribution using an 800 nm wavelength source on the breast phantom. The background absorption coefficient is $\mu_a = 0.0005$ mm^{-1}, with a reduced scattering coefficient $\mu_s' = 1.5742$. Glandular tissue ($\mu_a = 0.0059$ mm^{-1}, $\mu_s' = 1.12$ mm^{-1}) surrounds a tumor ($\mu_a = 0.0021$, $\mu_s' = 0.625$ mm^{-1}) located 32.5 mm deep from the right side of the phantom.

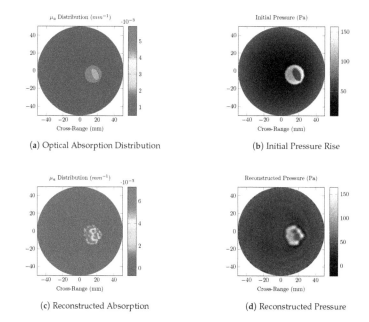

Figure 7. Photoacoustic computed tomography (PACT) simulations of the breast phantom at 800 nm wavelength; (**a**) the true optical absorption; (**b**) the initial pressure rise induced by the photoacoustic effect; (**d**) the reconstructed pressure, calculated using the time reversal algorithm; and (**c**) the reconstructed absorption, obtained by dividing the reconstructed pressure with the fluence. The elliptical region similar to the background is the tumor, surrounded by the glandular tissue.

3.4. RF-Induced Acoustic Computed Tomography (RACT) of the Breast Phantom

The radio frequency source for the breast phantom, similar to the homogeneous phantom, operates at 434 MHz. The field distribution is provided in Figure 8.

The forward simulation of the electric field is able to directly detect the tumor and an estimation of its location. The conductivity has a much more significant effect on the wave, since the wavelength is larger than the region of interest. In the large wavelength regime, scattering due to dielectric contrast does not distort the electromagnetic wave as much as the substantial conductivity. While the approximate location of the tumor can be inferred from the electric field intensity, the specific shape and glandular tissue identification requires further processing, such as the RF acoustic simulation. Figure 9 shows the initial pressure and the resulting reconstructed pressure and conductivity maps of the breast phantom.

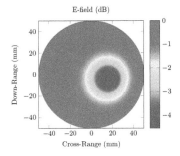

Figure 8. Total electric field intensity distribution of a 434 MHz radio frequency source inside the breast phantom. The background has a conductivity $\sigma = 0.0353$ S/m and dielectric constant $\varepsilon_r = 5.51$. Glandular tissue ($\sigma = 8.86$ S/m, $\varepsilon_r = 61.3$) surrounds a tumor ($\sigma = 13.03$ S/m, $\varepsilon_r = 25.25$) located 32.5 mm deep from the right side of the phantom.

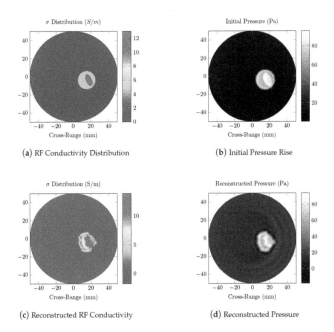

Figure 9. RF-induced acoustic computed tomography (RACT) simulations of the breast phantom using a 434 MHz RF source; (**a**) the true RF conductivity; (**b**) the initial pressure rise induced by the thermoacoustic effect; (**d**) the reconstructed pressure, calculated using the time-reversal algorithm; and (**c**) the reconstructed absorption, found by dividing the reconstructed pressure by the electric field intensity. The elliptical region with significantly larger conductivity is the tumor, surrounded by the glandular tissue.

4. Discussion

4.1. Merits of ONELAB for Thermoacoustic Imaging

Using an open source platform, ONELAB, allowed a single interface to simulate hybrid thermoacoustic imaging, with the choice of optical, radio-frequency, or other sources to induce a pressure wave response. The components of ONELAB Gmsh (mesh) and GetDP (FEM solver) were applied for this purpose. Since GetDP only solves the user defined finite element equations, fine control over every aspect of the thermoacoustic simulation was possible. This is especially important when performing reconstruction, since, for instance, time-reversal can have drastically different performance depending on the amount of time steps and the duration per time step. The benefits of Gmsh include its standard mesh creation, with the ability to automatically interface with other standard mesh formats. This allows the user to write a wrapper around any given finite element mesh. Besides simulating expected results, this platform is also ideal for post-processing the results. All of the data from the simulated PACT and RACT experiments was in the same format, used the same finite element mesh, without the need for interpolation or data transformation. Using the data in such a complementary fashion demonstrated the benefits of ONELAB over more established simulation tools used for a single imaging modality.

4.2. Simulation Workflow for both PACT and RACT

We first tested our algorithms on a simple phantom that consisted of two absorbing regions of different shapes inside a homogeneous background. The second phantom simulated a human breast with a tumor. The same phantom mesh was used to define both optical and RF-acoustic properties of the tissue. The optical diffusion equation was solved to obtain the optical fluence maps of the phantoms (Figures 2 and 6), while the solution to Maxwell's equations mapped the electric field distribution for these phantoms (Figures 4 and 8). Subsequently, maps of the initial pressure rises, due to electromagnetic absorption and the thermoacoustic effect, were generated using the optical fluence maps in PACT and electric field intensity in RACT. The initial pressure distributions of both PACT and RACT were then propagated using the same time-domain equations. Respective maps of the reconstructed pressure were generated using a time-reversal algorithm. Optical absorption and conductivity of the tissue phantoms were recovered by dividing the related reconstruction pressures by the optical fluence and electric field intensity distributions, respectively (Figures 3, 5, 7 and 9).

4.3. Analysis of PACT and RACT Results

Overall, our simulation results on two different tissue phantoms have shown that ONELAB can effectively simulate photoacoustic computed tomography (PACT) as well as RF-induced acoustic computed tomography (RACT). The optical fluence distribution (Figure 2) and electric field intensity (Figure 4) for the homogeneous phantom were significantly different, though the PACT/RACT reconstruction (Figures 3 and 5) is of similar accuracy. The relative error of the maximum absorption coefficient was 1.25% for the PACT case, and 8.70% in the RACT case. PACT for this phantom had a much lower relative error, since the contrast between absorber was much larger than the same difference in the RACT case. Imperfect reconstruction of the absorbers is due to error in reconstructing the initial pressure, instead of differences in the field intensity distribution. The relative shape of the absorbers, using either PACT or RACT, is easily identified from the reconstruction, with further processing only necessary for very precise characterizations. The successfully simulated results on a generic phantom gave us confidence to further validate our algorithms on a real tissue phantom mimicking the optical and RF properties of a human breast.

When the approximate optical and RF parameters corresponding to breast tissue are used (Figures 6 and 8), the optical and electric field intensities resemble each other, and have similar accuracy to the homogeneous phantom. The relative error of the maximum absorption coefficient was 23.2% for PACT and 6.92% for RACT. In the realistic tissue, RACT had more contrast, leading to less error in the reconstruction. The difference in field intensity between the homogeneous and breast phantom did not have a significant effect on the reconstruction accuracy. Both PACT and RACT are able to approximately reconstruct the tumor and the surrounding glandular tissue with no further processing. Even though the tumor had properties similar to the background, the PACT reconstruction was still able to identify the tumor surrounded by the glandular tissue. In RACT, the tumor is readily identified as being significantly different from the background.

In summary, our work demonstrated that ONELAB is a viable simulation platform for use in PACT and RACT, and is well suited for experiments that exploit both modalities. Reconstructed images of the phantoms provided both qualitative and quantitative information about the tissue optical and conductivity properties, including the size, shape, and location of the target regions. This work laid a foundation for future studies to develop and validate more robust multimodality reconstruction algorithms that will help improve quantitative accuracy.

5. Conclusions

This study has demonstrated the use of the tools Gmsh and GetDP, known together as ONELAB, as a single simulation platform for modeling both optical, as well as RF-induced, thermoacoustic computed tomography, i.e. PACT and RACT, respectively. To achieve PACT and RACT results, the propagation of optical, radio frequency, and acoustic waves were effectively modeled using solutions of the optical diffusion equation, Maxwell's equations, and time-domain wave equations. We validated our PACT and RACT algorithms using two types of tissue mimicking phantoms: a homogeneous phantom consisting of two absorbing targets and a breast phantom consisting of a tumor, with pre-defined optical and RF properties. Our results demonstrated that the optical and RF absorption properties of the respective tissue phantoms were accurately reconstructed using the proposed dual-modality computed tomography simulations in ONELAB. The use of the ONELAB software package allows for accurate multimodal thermoacoustic modeling, in order to develop and validate more robust algorithms for functional imaging of the human body.

Author Contributions: C.F. designed the simulations, analyzed the results, and wrote the paper. S.R.K. assisted in both the analysis of the results and writing of the paper.

Funding: The authors would like to thank the NIH for funding this work under NIH-NIBIB 4R00EB017729 (Sri-Rajasekhar Kothapalli) and PennState Hershey Center Institute Cancer Startup Funds (Sri-Rajasekhar Kothapalli).

Conflicts of Interest: The authors declare no conflict of interest.

Abbreviations

The following abbreviations are used in this manuscript:

ONELAB	Open Numerical Engineering Laboratory
GetDP	General Environment for the Treatment of Discrete Problems
RF	radio frequency
CT	computed tomography
PET	positron emission tomography
MRI	magnetic resonance imaging
UCT	ultrasound computed tomography
PACT	photoacoustic computed tomography
RACT	radio frequency acoustic computed tomography
TACT	thermoacoustic computed tomography
FEM	finite element method
FDM	finite difference method
PDE	partial differential equation
CAD	computer-aided design

IGES   initial graphics exchange specification
STL   stereolithography
DICOM   Digital Imaging and Communications in Medicine
NIR   near infrared
TM   transverse magnetic
PML   perfectly matched layer

References

1. Glick, S.J. Breast CT. *Annu. Rev. Biomed. Eng.* **2007**, *9*, 501–526. [CrossRef] [PubMed]
2. Vercher-Conejero, J.L.; Pelegrí-Martinez, L.; Lopez-Aznar, D.; Cózar-Santiago, M.D.P. Positron Emission Tomography in Breast Cancer. *Diagnostics* **2015**, *5*, 61–83. [CrossRef] [PubMed]
3. Lehman, C.D.; Schnall, M.D. Imaging in breast cancer: Magnetic resonance imaging. *Breast Cancer Res.* **2005**, *7*, 215–219. [CrossRef] [PubMed]
4. Boone, J.M.; Yang, K.; Burkett, G.W.; Packard, N.J.; Huang, S.Y.; Bowen, S.; Badawi, R.D.; Lindfors, K.K. An X-ray Computed Tomography/Positron Emission Tomography System Designed Specifically for Breast Imaging. *Technol. Cancer Res. Treat.* **2010**, *9*, 29–44. [CrossRef] [PubMed]
5. Martí-Bonmatí, L.; Sopena, R.; Bartumeus, P.; Sopena, P. Multimodality imaging techniques. *Contrast Media Mol. Imaging* **2010**, *5*, 180–189. [CrossRef] [PubMed]
6. Torigian, D.A.; Zaidi, H.; Kwee, T.C.; Saboury, B.; Udupa, J.K.; Cho, Z.H.; Alavi, A. PET/MR Imaging: Technical Aspects and Potential Clinical Applications. *Radiology* **2013**, *267*, 26–44. [CrossRef] [PubMed]
7. Sehgal, C.M.; Weinstein, S.P.; Arger, P.H.; Conant, E.F. A review of breast ultrasound. *J. Mammary Gland Biol. Neoplasia* **2006**, *11*, 113–123. [CrossRef] [PubMed]
8. Ku, G.; Wang, L.V. Scanning thermoacoustic tomography in biological tissue. *Med. Phys.* **2000**, *27*, 1195–1202. [CrossRef] [PubMed]
9. Xu, M.; Wang, L.V. RF-induced thermoacoustic tomography. In Proceedings of the Second Joint 24th Annual Conference and the Annual Fall Meeting of the Biomedical Engineering Society Engineering in Medicine and Biology, Houston, TX, USA, 23–26 October 2002; Volume 2, pp. 1211–1212. [CrossRef]
10. Wang, X.; Pang, Y.; Ku, G.; Xie, X.; Stoica, G.; Wang, L.V. Noninvasive laser-induced photoacoustic tomography for structural and functional in vivo imaging of the brain. *Nat. Biotechnol.* **2003**, *21*, 803–806. [CrossRef] [PubMed]
11. Yang, X.; Wang, L.V. Monkey brain cortex imaging by photoacoustic tomography. *J. Biomed. Opt.* **2008**, *13*, 044009. [CrossRef] [PubMed]
12. Akhouayri, H.; Bergounioux, M.; Da, S.A.; Elbau, P.; Litman, A.; Mindrinos, L. Quantitative thermoacoustic tomography with microwaves sources. *J. Inverse Ill-Posed Probl.* **2016**, *25*, 703–717. [CrossRef]
13. Pramanik, M.; Ku, G.; Li, C.; Wang, L.V. Design and evaluation of a novel breast cancer detection system combining both thermoacoustic (TA) and photoacoustic (PA) tomography. *Med. Phys.* **2008**, *35*, 2218–2223. [CrossRef] [PubMed]
14. Razansky, D.; Kellnberger, S.; Ntziachristos, V. Near-field radiofrequency thermoacoustic tomography with impulse excitation. *Med. Phys.* **2010**, *37*, 4602–4607. [CrossRef] [PubMed]
15. Heijblom, M.; Piras, D.; Brinkhuis, M.; van Hespen, J.C.G.; van den Engh, F.M.; van der Schaaf, M.; Klaase, J.M.; van Leeuwen, T.G.; Steenbergen, W.; Manohar, S. Photoacoustic image patterns of breast carcinoma and comparisons with Magnetic Resonance Imaging and vascular stained histopathology. *Sci. Rep.* **2015**, *5*, 11778. [CrossRef] [PubMed]
16. Emerson, J.F.; Chang, D.B.; McNaughton, S.; Jeong, J.S.; Shung, K.K.; Cerwin, S.A. Electromagnetic Acoustic Imaging. *IEEE Trans. Ultrason. Ferroelectr. Freq. Control* **2013**, *60*, 364–372. [CrossRef] [PubMed]
17. Cui, H.; Yang, X. In vivo imaging and treatment of solid tumor using integrated photoacoustic imaging and high intensity focused ultrasound system. *Med. Phys.* **2010**, *37*, 4777–4781. [CrossRef] [PubMed]
18. Dehghani, H.; Eames, M.E.; Yalavarthy, P.K.; Davis, S.C.; Srinivasan, S.; Carpenter, C.M.; Pogue, B.W.; Paulsen, K.D. Near infrared optical tomography using NIRFAST: Algorithm for numerical model and image reconstruction. *Commun. Numer. Methods Eng.* **2008**, *25*, 711–732. [CrossRef] [PubMed]
19. Treeby, B.E.; Cox, B.T. k-Wave: MATLAB toolbox for the simulation and reconstruction of photoacoustic wave fields. *J. Biomed. Opt.* **2010**, *15*, 021314. [CrossRef] [PubMed]

20. Wang, Z.; Ha, S.; Kim, K. Evaluation of finite element based simulation model of photoacoustics in biological tissues. In *Medical Imaging 2012: Ultrasonic Imaging, Tomography, and Therapy*; International Society for Optics and Photonics: San Diego, CA, USA, 2012; p. 83201L. [CrossRef]

21. Sowmiya, C.; Thittai, A.K. Simulation of photoacoustic tomography (PAT) system in COMSOL and comparison of two popular reconstruction techniques. In *Medical Imaging 2017: Biomedical Applications in Molecular, Structural, and Functional Imaging*; International Society for Optics and Photonics: Orlando, FL, USA, 2017; Volume 10137, p. 101371O. [CrossRef]

22. Geuzaine, C.; Remacle, J.F. Gmsh: A 3-D finite element mesh generator with built-in pre- and post-processing facilities. *Int. J. Numer. Methods Eng.* **2009**, *79*, 1309–1331. [CrossRef]

23. Dular, P.; Geuzaine, C.; Henrotte, F.; Legros, W. A general environment for the treatment of discrete problems and its application to the finite element method. *IEEE Trans. Magn.* **1998**, *34*, 3395–3398. [CrossRef]

24. Geuzaine, C.; Henrotte, F.; Remacle, J.F.; Marchandise, E.; Sabariego, R. ONELAB: Open Numerical Engineering LABoratory. In Proceedings of the 11e Colloque National en Calcul des Structures, Giens, France, 13–17 May 2013.

25. Treeby, B.E.; Zhang, E.Z.; Cox, B.T. Photoacoustic tomography in absorbing acoustic media using time reversal. *Inverse Probl.* **2010**, *26*, 115003. [CrossRef]

26. Cox, B.T.; Treeby, B.E. Artifact Trapping During Time Reversal Photoacoustic Imaging for Acoustically Heterogeneous Media. *IEEE Trans. Med. Imaging* **2010**, *29*, 387–396. [CrossRef] [PubMed]

27. Xu, M.; Wang, L.V. Universal back-projection algorithm for photoacoustic computed tomography. *Phys. Rev. E* **2005**, *71*, 016706. [CrossRef] [PubMed]

28. Popov, D.A.; Sushko, D.V. Image Restoration in Optical-Acoustic Tomography. *Probl. Inf. Trans.* **2004**, *40*, 254–278. [CrossRef]

29. Kuchment, P.; Kunyansky, L. Mathematics of thermoacoustic tomography. *Eur. J. Appl. Math.* **2008**, *19*, 191–224. [CrossRef]

30. Wang, L.; Jacques, S.L.; Zheng, L. MCML—Monte Carlo modeling of light transport in multi-layered tissues. *Comput. Methods Programs Biomed.* **1995**, *47*, 131–146. [CrossRef]

31. Kim, C.; Li, Y.; Wang, L.V. The study of quantitative optical absorption imaging by using Monte Carlo simulation of combined photoacoustic tomography and ultrasound-modulated optical tomography. In *Photons Plus Ultrasound: Imaging and Sensing 2012*; International Society for Optics and Photonics: San Francisco, CA, USA, 2012; Volume 8223, p. 82232C. [CrossRef]

32. Ammari, H. (Ed.) *Mathematical Modeling in Biomedical Imaging II: Optical, Ultrasound, and Opto-Acoustic Tomographies*; Mathematical Biosciences Subseries; Springer: Berlin/Heidelberg, Germay, 2012.

33. Wang, L.V. (Ed.) *Photoacoustic Imaging and Spectroscopy*, 1st ed.; CRC Press: Boca Raton, FL, USA, 2009.

34. Berenger, J.P. A perfectly matched layer for the absorption of electromagnetic waves. *J. Comput. Phys.* **1994**, *114*, 185–200. [CrossRef]

35. Kellnberger, S.; Hajiaboli, A.; Razansky, D.; Ntziachristos, V. Near-field thermoacoustic tomography of small animals. *Phys. Med. Biol.* **2011**, *56*, 3433–3444. [CrossRef] [PubMed]

36. Roden, J.A.; Gedney, S.D. Convolution PML (CPML): An efficient FDTD implementation of the CFS–PML for arbitrary media. *Microw. Opt. Technol. Lett.* **2000**, *27*, 334–339. [CrossRef]

37. Kaltenbacher, B.; Kaltenbacher, M.; Sim, I. A modified and stable version of a perfectly matched layer technique for the 3-d second order wave equation in time domain with an application to aeroacoustics. *J. Comput. Phys.* **2013**, *235*, 407–422. [CrossRef] [PubMed]

38. Evans, L.C. *Partial Differential Equations: Second Edition*, 2nd ed.; American Mathematical Society: Providence, RI, USA, 2010.

39. Surowiec, A.J.; Stuchly, S.S.; Barr, J.R.; Swarup, A. Dielectric properties of breast carcinoma and the surrounding tissues. *IEEE Trans. Biomed. Eng.* **1988**, *35*, 257–263. [CrossRef] [PubMed]

40. Jacques, S.L. Optical properties of biological tissues: A review. *Phys. Med. Biol.* **2013**, *58*, R37. [CrossRef] [PubMed]

41. Prahl, Scott. Assorted Spectra. Available online: https://omlc.org/spectra/index.html (accessed on 28 June 2018).

42. Lloyd, Bryn A. Dielectric Properties. Available online: https://itis.swiss/virtual-population/tissue-properties/database/dielectric-properties/ (accessed on 19 June 2018).

43. Gabriel, C. *Compilation of the Dielectric Properties of Body Tissues at RF and Microwave Frequencies*; Air Force Materiel Command, Brooks Air Force Base: San Antonio, TX, USA, 1996; p. 272.

44. Gabriel, C.; Gabriel, S.; Corthout, E. The dielectric properties of biological tissues: I. Literature survey. *Phys. Med. Biol.* **1996**, *41*, 2231. [CrossRef] [PubMed]

MDPI

St. Alban-Anlage 66

4052 Basel

Switzerland

Tel. +41 61 683 77 34

Fax +41 61 302 89 18

www.mdpi.com

Applied Sciences Editorial Office

E-mail: applsci@mdpi.com

www.mdpi.com/journal/applsci

CPSIA information can be obtained
at www.ICGtesting.com
Printed in the USA
BVHW020850040321
601386BV00041B/1029

9 783039 436439